Ecological Modeling

W.E.G.
To Linda, Jennifer, Stephanie, David, Amanda, Jessica,
Steven, Jacob, Joseph, and Lindsey

T.M.S.
To Mom, Dad and Margie, Craig and Shawntel, Ashley, and John

William E. Grant & Todd M. Swannack

Ecological Systems Laboratory
Department of Wildlife and Fisheries Sciences
Texas A&M University
College Station
Texas, USA

Ecological Modeling:

A common-sense approach to theory and practice

Blackwell Publishing

BLACKWELL PUBLISHING
350 Main Street, Malden, MA 02148–5020, USA
9600 Garsington Road, Oxford OX4 2DQ, UK
550 Swanston Street, Carlton, Victoria 3053, Australia

First published 2008 by Blackwell Publishing Ltd

1   2008

ISBN-13: 978-1-4051-6168-8

Library of Congress Cataloging-in-Publication Data

Grant, William E. (William Edward), 1947–
    Ecological modeling : a common-sense approach to theory and practice / William E.
Grant & Todd M. Swannack.
        p. cm.
    Includes bibliographical references and index.
    ISBN 978-1-4051-6168-8 (pbk. : alk. paper)   1. Ecology–Mathematical
models.   I. Swannack, Todd M.   II. Title.

    QH541.15.M3G735 2008
    577.01'5118–dc22

2007022239

A catalogue record for this title is available from the British Library.

Set in 11/13.5 pt Dante
by SNP Best-set Typesetter Ltd., Hong Kong

For further information on
Blackwell Publishing, visit our website at
www.blackwellpublishing.com

Contents

Part 2 Modeling theory

5 Theory III: model evaluation   79

6 Theory IV: model application   89

Part 3   Modeling practice

7 Some common pitfalls   93

Preface

Why did we write another book on modeling? Should you bother to read it? If you'll allow us to reflect a bit on the first question, you'll probably be able to answer the second.

The elder of us (WEG) has always wanted a simple textbook for the introductory systems modeling class he has been teaching in the Department of Wildlife and Fisheries Sciences at Texas A&M University since 1976. The first attempt to write such a text (Grant 1986) resulted from the fact that most modeling texts available in the 1970s and early 1980s were written for engineers, and ecological modeling books, including some landmark works, were primarily reference volumes rather than textbooks. A second attempt (Grant et al. 1997, 2001) a decade later resulted from a desire to have a textbook in Spanish for short courses that we (co-authors were Sandra Marín and Ellen Pedersen) had begun teaching in Latin America. During the process of writing the Spanish language text we found ourselves rewriting large portions of the text in English as well; a decade of student questions had brought new insights into relating theory to practice and advances in computers and software had provided interesting possibilities for giving readers a more hands-on introduction to modeling. This third attempt quite honestly was motivated initially by a desire to reduce costs to students via a shorter (paperback) version of the 1997 text and alleviation of ties to rather costly commercial software. But the ensuing scrutiny of the text from a new perspective, which emerged as a synthesis of another decade of reflection (WEG) and the innovative ideas of a new generation of ecological modelers (TMS), forced a reevaluation of what is essential to an introductory course. Independent of cost, the 1997 text now seems unnecessarily long and complex and the sophistication of associated software seems to distract attention from the simplicity of basic calculations.

Thus the present text is the result of a relentless culling of material that is not absolutely essential to the prudent development, evaluation, and use of systems models. We realize this reduction runs counter-current to the rapidly expanding range and sophistication of ecological modeling topics. We are in awe of the intellectual prowess of today's ecological modelers and of the stream of theoretical, analytical, and computational breakthroughs they produce (for example, see articles in the journal *Ecological Modelling*). But continuous innovation is a two-edged sword, potentially overwhelming beginners and, we would humbly suggest, blinding some experienced practitioners to basic errors hidden at the core of their sophisticated applications. Nowadays there is a wealth of books, journals, and websites that collectively describe virtually all aspects of ecological (biological, environmental, natural resource) modeling. What perhaps is less available, with all due respect to authors of current "Introduction to" modeling books, each with their particular strengths, is an introduction to the basic principles and practice of systems modeling within an ecological context. The question "What do I really need to know before I can build and use ecological models in a responsible manner?" is a critical one which hopefully is answered in the present text.

William E. Grant
Todd M. Swannack

References   Grant, W.E. 1986. *Systems Analysis and Simulation in Wildlife and Fisheries Sciences.* New York: Wiley.

Grant, W.E., E.K. Pedersen, and S.L. Marín. 1997. *Ecology and Natural Resource Management: systems analysis and simulation.* New York: Wiley.

Grant, W.E., S.L. Marín, and E.K. Pedersen. 2001. *Ecología y Manejo de Recursos Naturales: análisis de sistemas y simulación.* San José, Costa Rica: Instituto Interamericano de Cooperación para la Agricultura (IICA).

Acknowledgments

We would like to thank Keith Blount, Selma Glasscock, Nils Peterson, Ed Rykiel, Lloyd W. Towers III, and two anonymous reviewers for their comments on earlier versions of this book. We also would like to thank Linda Causey for drawing the box-and-arrow diagrams found throughout the text.

1
Introduction

The usefulness of ecological simulation modeling results as much from the process (problem specification, model development, and model evaluation) as from the product (the final model and simulations of system dynamics). Skill in the process of simulation modeling is gained primarily through (1) practice, (2) practice, and (3) practice. However, a keen awareness of what we are doing (in practice), why we are doing it (in theory), and why it makes (common) sense, is invaluable. Without this awareness we risk making silly, kindergarten-level, mistakes; even experienced modelers are not immune from these pitfalls, which often come hidden under a thick covering of sophisticated quantitative techniques and associated jargon. Thus, we have organized this book to emphasize the "oneness" of theory, practice, and common sense.

1.1 Common-sense solutions: three exercises

We begin in Chapter 2 with (1) practice, (2) practice, and (3) practice, in the form of three exercises. In each exercise, we are faced with a problem that requires us to project the dynamics of a particular system into the future under different scenarios of interest. The first deals with a group of hunter-gatherers harvesting food for the winter, the second with a population that might go extinct, and the third with management of a common pasture in which neighbors graze their animals. We first work through each problem in an informal, commonsensical way. We then present a short overview of the systems approach to problem solving, and briefly revisit the three problems from the systems perspective.

What should be obvious from these three examples is that projecting the dynamics of even relatively simple systems for which we have a good understanding and a solid database is not necessarily an easy matter. Apparently simple systems may exhibit surprisingly complex behavior; an understanding of the behavior of each part of the system does not guarantee an understanding of the behavior of the whole system. Attempts to deal with complex problems in a narrow or fragmentary way often lead to poor research design and ultimately to poor management decisions. We need an effective way of dealing with the complexity generated by interaction among the parts.

1.2 Modeling theory

We then take a more formal look at the simulation modeling process from the "systems perspective," describing development of the conceptual model (Chapter 3), quantification of the model (Chapter 4), evaluation of the model (Chapter 5), and application of the model (Chapter 6). The systems perspective, or systems approach, is both a philosophical perspective and a collection of techniques, including simulation, which emphasizes a holistic approach to problem solving as well as the use of mathematical models to identify and simulate important characteristics of complex systems. In the simplest sense, a system is any set of objects that interact, and a mathematical model is a set of equations that describes the interrelationships among these objects. By solving the equations comprising a mathematical model we can mimic, or simulate, the dynamic (time-varying) behavior of the system.

The basic approach is to (1) develop a conceptual model (box and arrow diagram) identifying specific cause–effect relationships among important components of the system in which we are interested, (2) quantify (write mathematical equations for) these relationships based on analysis of the best information available, (3) evaluate the usefulness of the model in terms of its ability to simulate system behavior under known scenarios and under an appropriately broad range of future scenarios, and (4) apply the model (conduct simulated experiments) to address our questions concerning system behavior under future scenarios.

1.3 Modeling practice

We next take a look at the practical application of simulation modeling, pointing out some of the pitfalls commonly encountered during model development (Chapter 7), and suggesting a strategy that we have found helpful in at least reducing the number of pits into which we fall (Chapter 8). Although theoretically it is convenient to describe the modeling process as proceeding smoothly through the four phases noted above, in practice we usually cycle through these phases several times. We seldom quantify the entire conceptual model before running simulations and evaluating model behavior. Rather, we usually construct a simple "running" model as quickly as possible and then expand it gradually through a series of small additions until we have quantified the entire model.

Pitfalls that we hope to avoid via this iterative approach range from the seemingly trivial, like failing to adequately define model objectives, to the foolhardy, like trying to track down numerical errors amongst tens or hundreds of interrelated equations that we (i.e., the computer) are solving for the first time. Common pitfalls include inappropriately bounding the system-of-interest, often due to inclusion of excessive detail in model structure, and underestimating the importance of time lags and negative feedback, often due to the erroneous idea that cause and effect must be tightly linked in time and space.

1.4 Theory, practice, and common sense

Having emphasized the oneness of "theory," "practice," and "common sense" as the organizational paradigm for this book, we feel obliged, perhaps somewhat ironically, to clarify the distinction we make among these three terms. By theory (Chapters 3–6), we do not refer to "high theory" in the sense of General System Theory or the Theory of Quantum Mechanics, but simply to four general activities that are viewed as essential to the development and application of any systems simulation model. By practice (Chapters 7 and 8), we refer to the practical application of these four general activities in a manner that experience suggests helps us avoid some common modeling pitfalls. By common sense, we refer to a logical, straightforward approach to problem solving, whether described informally or formally.

Thus, viewed on a continuum from "high theory" to "day-to-day practice," virtually all we present in this book is very practically oriented. Our goal is two-fold:

1 To show that the theory and the formal practice of ecological modeling blend seamlessly into a commonsensical approach to problem solving.
2 To demonstrate that the added rigor provided by a keen awareness of what we are doing, why we are doing it, and why it makes sense aids us greatly in dealing with dynamic systems whose complexity would otherwise by overwhelming.

1.5 Intended use of this book

We have written this book as a textbook for an introductory course in ecological modeling. We have relentlessly culled material that is not absolutely essential to the prudent development, evaluation, and use of systems models, with the goal of providing a useful answer to the question: "What do I really need to know before I can build and use eco-logical models in a useful, and responsible, manner?" We have empha-sized the simplicity of the modeling process, and we firmly believe that mastery of the basic principles of ecological modeling is well within the grasp of any ecologist or student of ecology. Proficiency comes with practice.

Does this mean that after mastering the material in this book one is ready to develop ecological models that can advance ecological theory or improve natural resource management? An analogous question might be: "Does this mean that after mastering a foreign language one is ready to become a foreign ambassador?" Obviously, in both cases "it depends." We opened this chapter by stating that the usefulness of ecological model-ing results as much from the process as from the models that result from that process. We believe anyone who deals with dynamic (time-varying) ecological systems can learn more about those systems via the modeling process as described in this book. Whether the benefits of that learning extend beyond the self-edification of the modeler depends on the ques-tions being addressed and the level of ecological expertise of the modeler, or team of modelers, who is addressing them. Whether it would be useful to employ sophisticated mathematical and computational (computer) techniques during the modeling process likewise depends on the question being addressed and the level of mathematical and computer program-ming expertise of the modelers. We should not confuse the specific eco-logical, mathematical, programming or other expertise needed to address a particular problem with the basic principles of systems modeling, which provide a general problem-solving approach within which to apply this specific subject-matter expertise.

We do not present alternative mathematical formats or computer code for the examples in this book. Nor do we include references to the primary ecological modeling literature in the body of the text. We felt this would distract us from our main theme. There is a rich literature describing diverse types of ecological (biological, environmental, natural resource) models developed using a variety of mathematical and computer pro-gramming formats. Much of this literature is readily available on the internet, and readers should have no difficulty finding examples in their particular area of interest. We do provide 60 references to scientific arti-cles containing ecological models in Appendix A. We intend these refer-ences to serve only as points of entry into the primary literature.

2

Common-sense solutions

In this chapter, we consider three problems, each of which requires us to project the dynamics of a particular system into the future under different scenarios of interest. The first deals with a group of hunter-gatherers harvesting food for the winter, the second with a population that might go extinct, and the third with management of a common pasture in which neighbors graze their animals. We first work through each problem in an informal, commonsensical way, using a common technique for problem solving (left column of Fig. 2.1). Next we present a short overview of the formal systems approach to problem solving. We revisit the three problems, describing the various steps in our common-sense solutions from the formal systems perspective, noting some modifications of the theory required in practice. In theory, problems are solved continuously from beginning to end. Practical applications, however, inevitably are cluttered with pitfalls, the avoidance of which requires certain compromises with theory. Many pitfalls encountered during the simulation modeling process are so common that we have found it useful to formalize a strategy for dealing with them. We are sure that similar strategies abound, but they typically are not shared widely, and rarely appear in print.

We will make reference to the three problems described in this chapter throughout the book as a means of demonstrating, in a commonsensical way, the practical application of simulation modeling theory. We have described the solutions in more detail than might seem warranted for such simple systems, although it should become clear that projecting the dynamics of even relatively simple systems is not necessarily a trivial task. We also have scattered numerous parenthetical entries throughout the solution descriptions. These entries will serve as points of reference to specific aspects of modeling theory and modeling practice that are presented in more detail later in the book. We recommend that you simply ignore these rather cryptic entries as you read through the exercises for the first time. However, lest they become a major distraction if left unexplained, we want to take a moment to define the symbols and put them into an appropriate context.

The symbols within parentheses refer to the specific aspects of modeling theory and modeling practice that are illustrated by a particular part of the common-sense solutions (Fig. 2.1). Roman numerals subscripted with lower-case letters (I_a, . . . , IV_c) refer to the theoretical phases of modeling and the steps within each phase, respectively (center column of boxes in Fig. 2.1). Capital italicized letters subscripted with a number or lower-case "i" (for i[th]) (CM, IDM, FM) refer to the practical activities that form our suggested strategy for simulation modeling (right column of boxes in Fig. 2.1). The essence of both of these formal systems approaches to problem solving might be summarized informally in quite simple terms (left column of boxes in Fig. 2.1). We first diagram the important relationships within our system-of-interest and guess the sort of system behavior these relationships might generate. We then begin to represent the relationships numerically, calculate the results of these numerical representations, and check to be sure that both the relationships and the results make sense. We advance little by little, relationship by relationship, with this quantify–calculate–check procedure, so that we can more easily identify the inevitable mistakes we make along the way. When we have the entire model quantified and checked, we calculate and interpret system behavior under appropriate scenarios to address our questions.

So, our intent is that these symbols will provide a useful cross-referencing of the steps involved in modeling theory and modeling practice, and will facilitate recognition of the commonality these formal problem-solving protocols share with the more informal, common-sense approach we first use to solve the three problems. However, once again, we recommend that you simply ignore the parenthetical entries as you read through the exercises for the first time.

2.1 Three problems

2.1.1 Harvesting food for the winter

PROBLEM: We want to know if recent deforestation will have a serious impact on a group of hunter-gatherers that live in the forest during winter. The group usually hunts in the savanna until the end of October and then moves to the forest to spend the winter.

BACKGROUND INFORMATION: Severe winter weather, which prevents hunting and gathering activities, typically arrives by the end of November. Thus, the group has only about a month to harvest the food items they need to survive the winter.

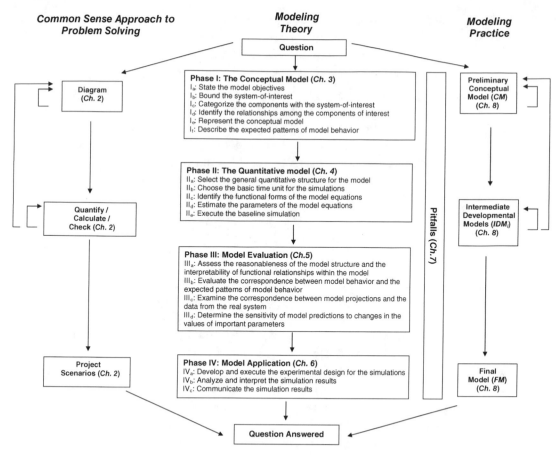

Fig. 2.1 Diagram cross-referencing a common sense approach to problem solving (left side) with the steps involved in modeling theory (center) and modeling practice (right side). Symbols refer to specific steps of modeling theory and practice (e.g., II$_a$, = step "a" of theoretical phase II, *IDM$_i$* = "ith" intermediate developmental model).

There usually are about 100 food items in the forest at the beginning of November. For all practical purposes, food items are a nonrenewable resource during November since plants no longer produce fruits and both birth and natural death rates of animal populations are essentially zero. Historically, the group has had no trouble harvesting the 75 food items they need to survive the winter, although the group has told us the rate at which they can harvest food items decreases noticeably as abundance of food items decreases. The group estimates they can harvest about 10% of the available food items per day. As a result of the recent deforestation, we estimate only about 80 food items will remain in the forest when the group arrives this fall. Although there still appears to be enough food available, we wonder if the group will have enough time to harvest all they need before the severe weather arrives.

Diagram To help answer this question, we decide to calculate food harvest and the accumulation of food items by the hunter-gatherers over a 30-day period, beginning with 80 food items in the forest (I_a) (CM). We decide to sketch a diagram to help us visualize how the hunter-gatherer system works. If we include the number of food items in the forest, the number of items harvested from the forest, and their accumulation for the winter, that should be everything we need to consider. The number of food items in the forest decreases, and the number of harvested food items increases, as items are harvested (Fig. 2.2a). The number of food items harvested is determined by the number of food items remaining in the forest (Fig. 2.2b) (I_b, I_c, I_d, I_e,)(CM).

Expected patterns Before we begin our calculations, we decide to take our best guess at how the number of food items in the forest might decrease, and the number of harvested food items might accumulate, assuming there are only 80 food items available at the beginning of November (Fig. 2.3b). We also sketch how food items historically have decreased, and accumulated, when there were 100 food items at the beginning of November (Fig. 2.3a) (I_f)(CM). These guesses should provide a good check on our calculations; if they are very different, it suggests we either made a mistake in our calculations or we guessed wrong, or both. In any event, we will be obliged to resolve the difference. If our calculations of the decrease in 100 food items differ from the manner in which we actually have observed them to decrease, we will know our calculations are flawed. If our calculations of the decrease in 80 food items differ from our guesses, we will need to recheck our calculations, reconsider our guesses, and decide how much confidence we will place on each.

Quantify, calculate, check Since the number of food items harvested depends on the current number of food items in the forest, which is continually

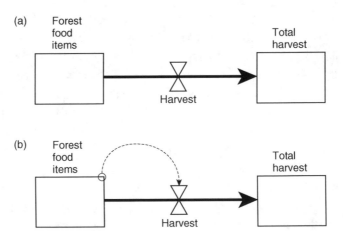

Fig. 2.2 Sequential development of the conceptual model representing food harvest by the group of hunter-gatherers in the forest during November.

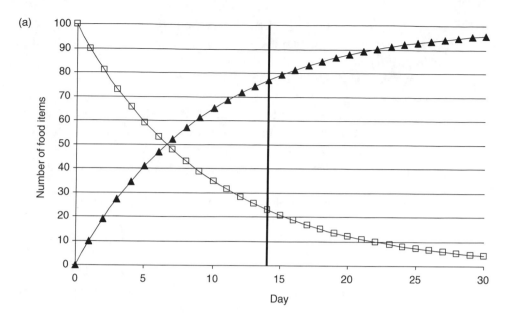

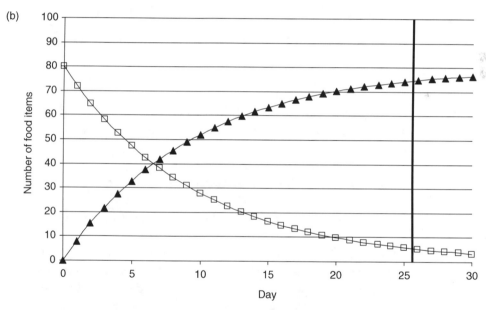

Fig. 2.3 (a) Our expectations of how the number of food items in the forest will decrease (open squares) and how many food items will be gathered (closed triangles) during November if we begin with 100 items. (b) Our expectations of how the number of food items in the forest will decrease (open squares) and how many food items will be gathered (closed triangles) during November if we begin with 80 items. The solid vertical line indicates the time it takes to gather the 75 items necessary to survive the winter.

changing, we decide to do a series of calculations, each representing the harvest of food items over a specific period of time $(II_a)(IDM_{last})$. We decide that recalculating harvest and updating food items in the forest each day should give us a good approximation, for our purposes, of changes in harvest rate $(\text{II}_b)(IDM_{last})$. Based on what the group has told us, we know they can harvest about 10% of the available food items per day $(\text{II}_c, \text{II}_d)(IDM_{last})$.

We now begin the calculations.* First, as a point of reference, we will calculate how the number of food items decrease during the 30-day period of November if we begin with 100 items $(\text{II}_e)(CM)(IDM_{last})$. When there are 100 food items in the forest, the group can harvest 10 items during the first day (10% of 100). Thus, the forest should contain 90 food items after 1 day (eq. 2.1) and total harvest should equal 10 (eq. 2.2).

$$\text{Amount of food in forest after the first day} = (100 - 10) = 90 \quad \text{(eq. 2.1)}$$

$$\text{Total harvest after the first day} = (0 + 10) = 10 \quad \text{(eq. 2.2)}$$

The harvest during the second day, which begins with 90 food items in the forest, is 9 items (10% of 90). Thus, the forest should contain 81 food items after 2 days (eq. 2.3) and the total amount harvested should equal 19 (eq. 2.4).

$$\text{Amount of food in forest after the second day} = (90 - 9) = 81 \quad \text{(eq. 2.3)}$$

$$\text{Total harvest after the second day} = (10 + 9) = 19 \quad \text{(eq. 2.4)}$$

For the third day, harvest rate is 8.1 items (10% of 81), and after the third day's harvest we calculate there would be 72.9 items in the forest (eq. 2.5) and a total of 27.1 items harvested (eq. 2.6).

$$\text{Amount of food in forest after the third day} = (81 - 8.1) = 72.9 \quad \text{(eq. 2.5)}$$

$$\text{Total harvest after the third day} = (19 + 8.1) = 27.1 \quad \text{(eq. 2.6)}$$

We repeat these calculations for days 4 through 30 and determine the number of food items in the forest will decrease to 4.24, and the number of items harvested will increase to 95.76, in 30 days. Total harvest surpasses the required 75 food items during day 14 (total harvest equals 77.12 items at the beginning of day 15).

Before calculating how the number of food items decrease if we begin with 80 items, we compare our calculations to the manner in which we

---

* Note that this problem is so simple that an analytical solution exists:
(Number of remaining food items at any time t) = (Initial number of food items) $\star$ $e^{-0.1 \star t}$, where e is the base of natural logarithms and t the number of days since harvesting began. Nonetheless, its simplicity serves us well to demonstrate the type of calculations involved in a numerical solution. We comment on analytical versus simulation models in Chapter 4 (Section 4.3.2).

actually have observed 100 items decrease during November. We also reevaluate the general form and logic of our calculations. We decide we have included the most important factors in our calculations in a reasonable manner $(III_a)(IDM_{last})$. Also, the results of our calculations correspond well both qualitatively $(III_b)(IDM_{last})$ and quantitatively $(III_c)(IDM_{last})$ with the manner in which we have observed the 100 items decrease (Fig. 2.3a).

Project scenarios We now are ready to calculate how the number of food items decrease if we begin with 80 items $(IV_a)(FM)$. When there are 80 food items in the forest, the group can harvest 8 items during the first day (10% of 80). Thus, we calculate the forest should contain 72 food items after one day (eq. 2.7), and total harvest should equal 8 (eq. 2.8).

$$\text{Amount of food in forest after the first day} = (80 - 8) = 72 \qquad (\text{eq. 2.7})$$

$$\text{Total harvest after the first day} = (0 + 8) = 8 \qquad (\text{eq. 2.8})$$

The harvest during the second day, which begins with 72 food items in the forest, is 7.2 items (10% of 72). Thus, we calculate the forest should contain 64.8 food items after 2 days (eq. 2.9), and total harvest should equal 15.2 (eq. 2.10).

$$\text{Amount of food in forest after the second day} = (72 - 7.2) = 64.8 \quad (\text{eq. 2.9})$$

$$\text{Total harvest after the second day} = (8 + 7.2) = 15.2 \quad (\text{eq. 2.10})$$

For the third day, harvest rate is 6.48 items (10% of 64.8), and we calculate 58.32 items are left in the forest (eq. 2.11) and a total of 21.68 items have been harvested (eq. 2.12).

$$\text{Amount of food in forest after the third day}$$
$$= (64.8 - 6.48) = 58.32 \qquad (\text{eq. 2.11})$$

$$\text{Total harvest after the third day} = (15.2 + 6.48) = 21.68 \ (\text{eq. 2.12})$$

We repeat these calculations for days 4 through 30 and determine the number of food items in the forest will decrease to 3.39, and the number of items harvested will increase to 76.61, in 30 days. Total harvest surpasses the required 75 food items during day 27 (total harvest equals 75.35 items at the beginning of day 28).

We now compare our calculations to our guesses of how the number of food items might decrease if we begin with 80 items $(IV_b)(FM)$. The results of our calculations correspond well both qualitatively $(IV_b)(FM)$

and quantitatively (IV$_b$)(*FM*) with the manner in which we guessed the 80 items would decrease (Fig. 2.3b). Thus, we decide our estimate of 27 days for the group of hunter-gatherers to harvest the 75 food items is a useful one.

Answer Based on the best information available, it looks like the group should have enough time to gather the food they require for winter. However, the group may have only a few (we estimated 3) "extra" days before the expected arrival of severe weather (IV$_c$)(*FM*).

2.1.2 Estimating the probability of population extinction

PROBLEM: We are concerned about the future of an animal population on a small, isolated island that is subject to hurricanes. We are particularly concerned because climatologists are projecting an increase in the frequency of occurrence of hurricanes over the next few decades. Historically, hurricanes have occurred about once every 10 years, although we do not have data indicating the exact years of occurrence.

BACKGROUND INFORMATION: Available data indicate population size on the island has fluctuated between 70 and 400 individuals over the past several decades. Data also indicate that annual per capita birth rate decreases linearly from a maximum of 0.7 when population size is 50 or lower to a minimum of 0.5 when population size reaches 400. Annual death rate usually is 50%, but increases to 99% during hurricane years. Fortunately, hurricanes do not affect annual per capita birth rate, since the young-of-the-year still are buried in the soil as eggs during hurricane season. Currently, as a result of a hurricane last year, only 100 individuals, the majority of whom survived the hurricane as eggs, remain on the island.

Solution Diagram To help consider the situation in more concrete terms, we decide to calculate probabilities of population extinction over the next 50 years assuming the current annual hurricane probability (10%), and assuming an annual probability of 50% (I$_a$)(*CM*). We decide to sketch a diagram, just as we did for our hunter-gatherer problem, to help us visualize how the system we are interested in works. If we include population size, births, deaths, birth rate, death rate, and hurricanes, that should cover everything we need to consider. Population size changes as a result of births and deaths (Fig. 2.4a), with births controlled by birth rate (per capita) and population size, and deaths controlled by death rate (per capita) and population size (Fig. 2.4b). Birth rate itself depends on population size (Fig. 2.4c), and death rate depends on hurricanes (Fig. 2.4d) (I$_b$, I$_c$, I$_d$, I$_e$)(*CM*).

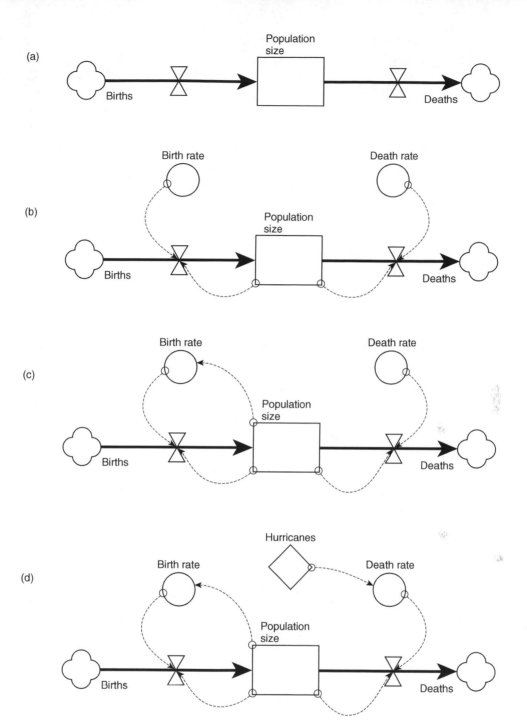

Fig. 2.4 Sequential development of the conceptual model representing the dynamics of the island population.

Expected patterns Before we begin our calculations, just as we did for our hunter-gatherer problem, we decide to describe the patterns we expect to see based on our current understanding of the population $(I_f)(CM)$. Since we know per capita birth rate decreases as population size increases, we expect, based on basic principles of population growth, the population will exhibit sigmoid (S-shaped) growth from a small size to carrying capacity (maximum size the island will support). We guess our population will grow from 100 to 400 individuals in about 25 years in the absence of hurricanes (Fig. 2.5a). If we superimpose the average historic frequency of hurricanes (once every 10 years), we expect the population will increase in a "two steps forward and one step backwards" fashion, and will not quite be able to reach 400 individuals before the next hurricane occurs (Fig. 2.5a). If we increase hurricane frequency to once every other year (50% annual probability), we expect the population will not be able to sustain itself (Fig. 2.5a). As in the hunter-gatherer example, these expected patterns should provide a good check on our calculations.

Quantify, calculate, check Since the rate of population change depends on its current value, which is continually changing, we decide to do a series of calculations, each representing the change over a specific period of time $(II_a)(IDM_1)$. We decide that recalculating and updating population size each year should provide adequate detail of temporal patterns for our purposes $(II_b)(IDM_1)$.

Now we are ready to begin our calculations. But since we are dealing with quite a few interacting components, we decide not to try to base the first projection on all of them. If we make an error or two, they might be hard to track down among so many interdependent calculations. Rather, we decide to begin with some very simple calculations, and then proceed step by step, adding a bit more complexity at each step. First we will calculate population growth assuming constant birth and death rates, next we will add calculations representing birth rate as a function of population size, and finally we will add calculations representing death rate as a function of hurricanes. We will check for potential errors at each step by comparing the results of our calculations to our expectations. Depending on the simplifying assumptions we have made for any given set of calculations, we may alter our expectations from those we described for the complete set of calculations.

We begin with calculations related to Fig. 2.4b, assuming no hurricanes, and constant birth and death rates. Since we have not yet included the density-dependent effects on birth rate, we expect our calculations will differ considerably from our expectations in Fig. 2.5a; without density-dependent birth rates, we expect J-shaped (exponential) rather than S-shaped (sigmoid) growth $(I_f)(CM)$. Based on Fig. 2.4b, we know we can calculate changes in population size by adding sequential birth and death

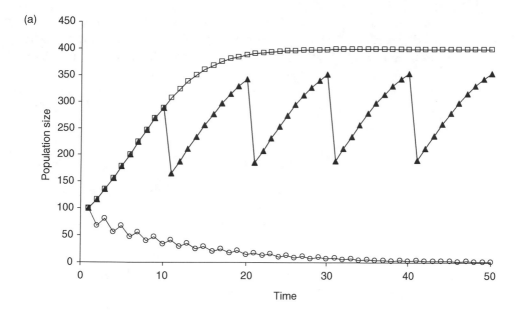

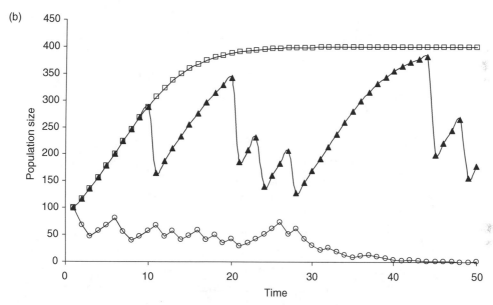

Fig. 2.5 (a) Our expectations of population dynamics without hurricanes (open squares), with the historic annual occurrence (10%) of hurricanes (closed triangles), and with an annual hurricane occurrence of 50% (open circles). (b) Population dynamics without hurricanes (open squares), with the historic annual probability (0.1) of hurricanes (closed triangles), and with an annual hurricane probability of 0.5 (open circles), according to our calculations. A single, typical, series of calculations (out of the 100 replicate series) representing each hurricane probability was chosen.

increments. We know annual death rate is 50% in nonhurricane years, and we will assume annual per capita birth rate remains constant at its maximum of 0.7 (II_c, II_d)(IDM_1). We decide to begin our calculations with a population of 100 individuals, which is the current size. We calculate births and deaths for the first year and add and subtract them, respectively, from the current population size to get the population size at the beginning of the second year (eqs 13–15).

$$\text{1st year births} = (0.7 \star 100) = 70 \qquad \text{(eq. 2.13)}$$

$$\text{1st year deaths} = (0.5 \star 100) = 50 \qquad \text{(eq. 2.14)}$$

$$\text{Population size} = (100 + 70 - 50) = 120 \qquad \text{(eq. 2.15)}$$

For the second year, we re-calculate births and deaths using the population size resulting from eq. 2.15, and use these results to calculate the population size after 2 years (eqs. 16–18).

$$\text{2nd year births} = (0.7 \star 120) = 84 \qquad \text{(eq. 2.16)}$$

$$\text{2nd year deaths} = (0.5 \star 120) = 60 \qquad \text{(eq. 2.17)}$$

$$\text{Population size} = (120 + 84 - 60) = 144 \qquad \text{(eq. 2.18)}$$

We repeat these calculations for subsequent years and determine the population is, indeed, growing exponentially, and will continue to do so forever. Although taken out of context these results obviously are unreasonable, within the context of our interim expectations of exponential growth, the calculations seem reasonable (III_a)(IDM_1), as do the results, both qualitatively (III_b)(IDM_1) and quantitatively (III_c)(IDM_1). Thus, we will proceed.

We now continue with calculations related to Fig. 2.4c, still assuming no hurricanes, but now with birth rates depending on population size. We will compare the results of these calculations to our expectations of population growth without hurricanes (Fig. 2.5a). This will serve as a point of reference for subsequently assessing the effect of hurricanes. We know annual per capita birth rate decreases linearly from a maximum of 0.7 when population size is 50 or fewer to a minimum of 0.5 when population size reaches 400 individuals (Fig. 2.6) (II_c, II_d)(IDM_2). We again decide to begin our calculations with a population of 100 individuals (II_e)(IDM_2). When population size is 100, annual birth rate (individuals/individual-year) is 0.67 (Fig. 2.6), and annual death rate (proportion of population dying/year) is 0.5. Thus we calculate the population should contain 117 individuals after 1 year (eq. 2.19).

$$\text{Population size} = [100 + (0.67 \star 100) - (0.5 \star 100)] = 117 \qquad \text{(eq. 2.19)}$$

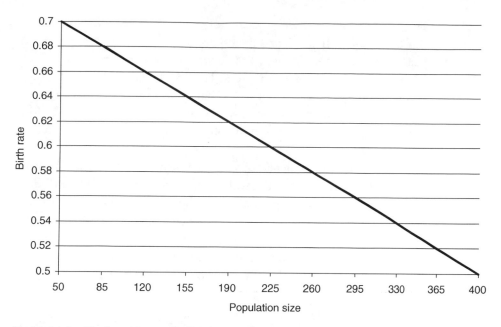

Fig. 2.6 Relationship of annual per capita birth rate (individuals born/individual in the population-year) to population size. Birth rate decreases linearly from 0.7 to 0.5 as population size increases from 50 to 400 for population sizes ≥50 and ≤400. Otherwise the birth rate is 0.7 for population size <50 and 0.5 for population size >400.

During the second year, the birth rate decreases due to an increased population size (Fig. 2.6), and when population size is 117, the birth rate is 0.662. The death rate remains 0.5. Thus we calculate the population should contain 135.9 individuals after 2 years (eq. 2.20).

$$\text{Population size} = [117 + (0.662 \star 117) - (0.5 \star 117)] = 135.9 \quad \text{(eq. 2.20)}$$

We repeat these calculations for subsequent years and determine the population should contain about 400 individuals in about 30 years. (We calculate that population size after 30 years is 399 and, since the population is approaching 400 ever more slowly, decide to terminate the calculations at that point.) We decide that our calculations seem reasonable $(III_a)(IDM_2)$, and our calculations correspond well both qualitatively $(III_b)(IDM_2)$ and quantitatively $(III_c)(IDM_2)$ with our expectations of sigmoid growth (Fig. 2.5a,b). Thus, we will proceed.

We continue with calculations related to Fig. 2.4d, with death rates now depending on the occurrence of hurricanes. We will compare the results of these calculations to our expectations of population growth with the historic frequency of hurricanes (Fig. 2.5a). Since we want to include the historic unpredictability of hurricanes explicitly in our calculations of death rate, we need to devise some way to "cause" a hurricane about every 10 years. We decide to put 1 black and 9 white marbles in a bowl, and, before each set of calculations, close our eyes and pick a marble out of the bowl. If we draw a black marble, a hurricane will occur that year and the death rate will be 0.99; if we draw a white marble, there will be no hurricane and the death rate will be 0.5 (II_c, II_d)(IDM_{last}). Thus, there will be a 10% chance (a 1 out of 10 chance of picking the black marble) of a hurricane occurring in any given year. (We need to put the marble we drew back in the bowl with the other marbles and mix them up each time.)

We now begin our calculations for the historic annual hurricane probability of 10%. Again we begin with a population size of 100 (II_e)(IDM_{last}). By chance, no hurricane occurs during the first year (we drew a white marble), thus, population size is again 117 after 1 year (eq. 2.19). By chance, a hurricane does occur during the second year (black marble), thus, the death rate becomes 0.99 and the birth rate is again 0.662 (Fig. 2.6), and we calculate the population should contain 78.6 individuals after 2 years (eq. 2.21):

$$\text{Population size} = [117 + (0.662 \star 117) - (0.99 \star 117)] = 78.6 \quad \text{(eq. 2.21)}$$

We repeat these calculations for years 3 through 50; by chance, hurricanes occur in years 16, 29, 42, and 43, and population size after 50 years is 275 (Table 2.1a). Since the occurrence of hurricanes almost surely will be different every time we perform our calculations, we decide to repeat the whole series of calculations 99 more times to be sure we have a representative sample (see Section 4.5.1) from which to estimate probability of population extinction. Thus, we will have 100 series of calculations from which to estimate the probability of extinction (Table 2.1b). The form and logic of these calculations seem reasonable (III_a)(IDM_{last}). Also, the results of our calculations are similar to our expectations in that population size fluctuates between approximately 70 and 400 individuals (Fig. 2.5a,b) (III_b, III_c)(IDM_{last}). We decide our calculations provide a useful representation of population dynamics and the probability of population extinction.

Project scenarios We now are ready to address our initial questions. That is, to calculate probabilities of population extinction over the next 50 years assuming the current annual hurricane probability (10%), and assuming an annual probability of 50% (IV_a)(FM). In this case, we already have calculated the dynamics of an initial 100-individual population for 50 years with an annual hurricane probability of 10%. We will repeat these calculations in the same manner 100 more times, but assuming an annual hurricane probability of 50% (we put 5 black and 5 white marbles in the bowl).

Results of our calculations suggest probabilities of population extinction during the next 50 years are less than 1% and about 60%, and average population sizes after 50 years are about 289 and 4, with annual hurricane

Table 2.1 (a) Example set of calculations of population dynamics assuming an annual hurricane probability of 10%. (b) Summary of results from 100 sets of calculations (100 simulations) assuming annual hurricane probabilities of 0.1 and 0.5. (c) Summary of results from 100 sets of calculations (100 simulations) assuming annual hurricane probabilities of 0.2–0.4. Probabilities of extinction were calculated as the number of times (out of 100 sets of calculations) that final population size was less than 2.

Table 2.1a

Year	Birth rate	Births	Population size	Death rate	Deaths	Hurricane
1	0.671429	67.14	100	0.5	50	No
2	0.661633	77.51	117.143	0.99	115.97	Yes
3	0.683613	53.78	78.677	0.5	39.34	No
. . .						
15	0.547933	173.21	316.117	0.5	158.06	No
16	0.539275	178.65	331.269	0.99	327.96	Yes
17	0.624596	113.65	181.958	0.5	90.98	No
. . .						
28	0.517752	191.02	368.935	0.5	184.47	No
29	0.514009	193	375.484	0.99	371.73	Yes
30	0.616139	121.23	196.757	0.5	98.38	No
. . .						
41	0.515211	192.37	373.381	0.5	186.69	No
42	0.511965	194.07	379.061	0.99	375.27	Yes
43	0.615511	121.78	197.857	0.99	195.88	Yes
44	0.657851	81.42	123.761	0.5	61.88	No
. . .						
49	0.595996	138.28	232.008	0.5	116	No
50	0.583269	148.31	254.279	0.5	127.14	No
Final			275.453			

Table 2.1b

Hurricane probability	Mean population size	Min. population size	Max. population size	Probability of extinction
0.10	288.91	82	399.99	<1.0
0.50	3.98	0.01	40.41	0.59

Table 2.1c

Hurricane probability	Mean population size	Min. population size	Max. population size	Probability of extinction
0.20	180.03	54.91	395.89	<1.0
0.30	96.48	3.25	392.69	<1.0
0.40	30.25	0.26	239.23	0.07

probabilities of 10% and 50%, respectively (Table 2.1b) $(IV_b)(FM)$. We decide these results seem reasonable when compared to our initial expectations (Fig. 2.5a,b), and the logic of our calculations still appears to be sound.

Answer Based on these results, it looks like there may be cause for concern about the future of the animal population if hurricanes become five times more likely to occur.

Further questions/answers These results answer our initial questions (IV$_b$)(*FM*), but, as often happens, raise some others. Since the probability of extinction is extremely low with the 10% hurricane probability, and dangerously high with the 50% probability, we decide to perform some additional calculations with several intermediate hurricane probabilities. We want to get a better idea of the hurricane probabilities at which the probability of extinction begins to approach dangerously high levels. Results of these additional calculations suggest that if annual hurricane probability approaches 40%, the population will be in considerable danger (about a 7% chance of extinction, with an average final population size of about 30) (Table 2.1c).

2.1.3 Managing the Commons

PROBLEM: Our neighbor has just informed us he plans to begin increasing the number of animals he puts on a common pasture we share. In the past, we both have put one animal on the Commons at the beginning of January and have removed it at the end of December. Apparently our neighbor has found someone who will buy whatever animals he produces as a single unit based on total herd biomass, and he plans to increase the size of his herd by 1 animal per year for the foreseeable future. On the one hand, it seems like we also should add animals to our "herd" so we take advantage of our fair share of the available forage. On the other hand, we are a bit worried about overgrazing. (Similarity to the classic scenario of Hardin (1968) is intended.)

BACKGROUND INFORMATION: We have collected a rather large body of information about the Commons, and similar pastures, over the years. We know that forage on the Commons fluctuates seasonally, but usually stays pretty close to 1000 biomass units. Rate of forage growth depends on the current biomass of forage. Forage growth rate decreases linearly from a maximum of 1 unit of biomass produced per unit of forage biomass per month to 0 as forage biomass increases from 200 to 1200 units. Forage is lost due to grazing by the animals. Each animal requires 1 unit of forage biomass per unit of animal biomass per month. There also is a nongrazing forage loss due to trampling, which is negligible with only 2 animals on the Commons. However, we suspect this loss may become important as the number of animals increases. We guess that each animal might trample about 2% of the forage each month.

We also know that animals are 2 months old and weigh about 10 biomass units when they are put on the Commons, and weigh about 90 biomass units when they are removed at 14 months of age. The maximum rate at which the animals can

gain weight, assuming there is enough forage to meet all of their forage requirements, depends on their current weight. Monthly weight gain decreases linearly
from a maximum of 1.25 times current body weight to 0 as weight increases from
5 (the lower lethal body weight) to 100 units. Actual weight gain depends on the
relative availability of forage. As the ratio of forage biomass available to forage
biomass required decreases from 5 to 1, the proportion of maximum possible
weight gain realized decreases from 1 to 0. Grazing animals normally metabolize
10% of their current biomass each month due to energy expenditures to meet
maintenance requirements. However, energy expenditures of females increase by
50% during the 6 months they are pregnant, and by 100% during the 3 months
they are nursing their young. Females become sexually mature at 18–20 months
of age and usually have two young per litter. Finally, we have data from a grazing
experiment conducted on three pastures similar to the Commons (Fig. 2.7).

Diagram To help resolve this dilemma, we decide to make some projec- *Solution*
tions of forage biomass and animal weights assuming (1) we increase the
number of animals we put on the Commons at the same rate our neighbor does, and (2) we continue to put just 1 animal on the Commons
each year. Specifically, we would like to know (1) if forage biomass will

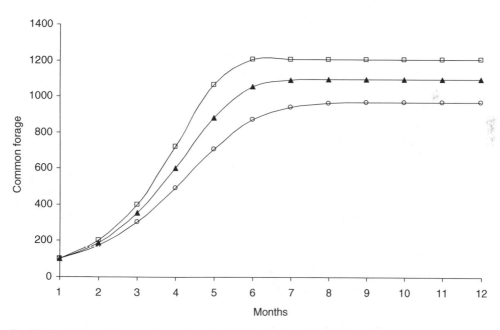

Fig. 2.7 Results from a 1-year grazing experiment conducted on three pastures similar to the
Commons. Initial forage biomass in each pasture was 100 biomass units, and the pastures contained
0 (open squares), 1 (closed triangles), and 2 (open circles) grazing animals, respectively, throughout
the year.

decrease and (2) if the final weights, at harvest, of our animals will decrease over the next 10 years (I_a)(*CM*). We decide to sketch a diagram to help us visualize how the system we are interested in works. Although the system-of-interest related to the Commons seems (and is!) much more complex than those related to our hunter-gatherer and population problems, the general idea is the same. We want to represent only the components we are interested in projecting into the future and the most important processes that affect them. We will want to include the forage biomass on the Commons (Fig. 2.8), the weight of our animals (Fig. 2.9), and the grazing and nongrazing processes that link the two and cause them to change over time (Fig. 2.10) (I_b, I_c, I_d, I_e)(*CM*). Let's take a step-by-step look at how we might put all the pieces together.

Regarding forage biomass, grass grows and biomass accumulates on the Commons (Fig. 2.8a). But forage biomass does not increase forever, even if we do not put animals on the Commons. Forage growth rate decreases as forage accumulates (Fig. 2.8b). If we do put animals on the Commons, there are two losses of forage biomass, one due to grazing (Fig. 2.8c) and one due to trampling (nongrazing loss) (Fig. 2.8d). Regarding weight of our animals, animals eat and gain weight (Fig. 2.9a). But the rate of weight gain decreases as animal weight increases (Fig. 2.9b). Animals also may lose weight if they can not eat enough to cover their maintenance requirements (Fig. 2.9c). Regarding the linkage of forage and animal weight dynamics, the cause–effect relationships are reciprocal. Each animal causes both grazing and nongrazing losses of forage (Fig. 2.10a), and weight gain of each animal depends on relative availability of forage (Fig. 2.10b). Since we have decided to represent forage biomass for the entire Commons, and animal weight on an individual-animal basis (average animal weight), we will need to convert the grazing and non-grazing losses of forage caused by each animal to an entire-herd basis (Fig. 2.10a). We also will need to represent relative forage availability on a per capita basis to affect weight gain of individual animals (Fig. 2.10b).

Expected patterns Before we calculate our projections, just as we did for our hunter-gatherer and population problems, we decide to describe the patterns we expect to see based on our current understanding of the Commons (I_f)(*CM*). Based on our personal observations, we expect that, with 2 animals on the Commons, forage biomass will fluctuate seasonally, but stay pretty close to 1000 biomass units year after year (Fig. 2.11). If each year we increase by 2 the number of animals on the Commons, we guess that forage biomass will decrease markedly over time (Fig. 2.11). If each year we increase by one the number of animals on the Commons, we guess that forage biomass will decrease roughly half as fast as increasing by 2 animals per year (Fig. 2.11). Also, we guess that, after 10 years

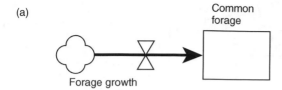

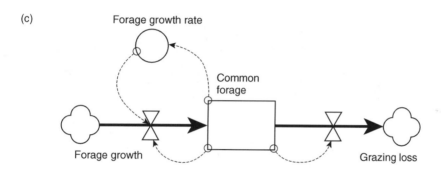

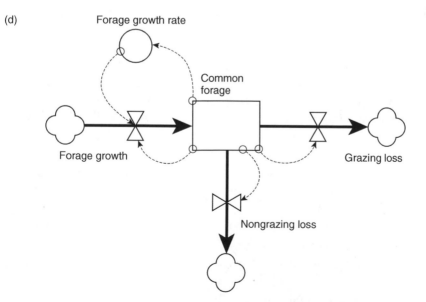

Fig. 2.8 Sequential development of the conceptual model representing forage dynamics on the Commons.

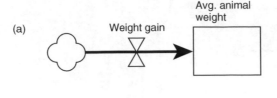

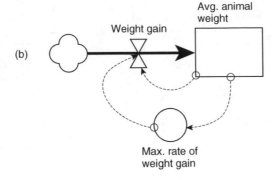

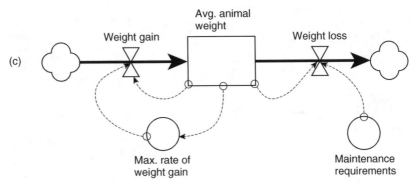

Fig. 2.9 Sequential development of the conceptual model representing animal weight dynamics on the Commons.

of increasing by 2 the number of animals on the Commons, harvest weights of animals will have decreased from the current 90 weight units to 20. After 10 years of increasing by 1 the number of animals on the Commons, we guess harvest weights will be around 45. Finally, we guess that under all three stocking scenarios animal weight will increase continually, in a linear fashion, from the initial 10 units to the harvest weight. As in the two previous examples, these expected patterns should provide a good check on our calculations.

Quantify, calculate, check Since the rates at which both forage biomass and animal weight change depend on their current values, which are continually changing, we decide to do a series of calculations, each representing the changes over a specific period of time (II$_a$)(IDM$_1$). We decide that recalculating and updating both forage biomass and animal weight each

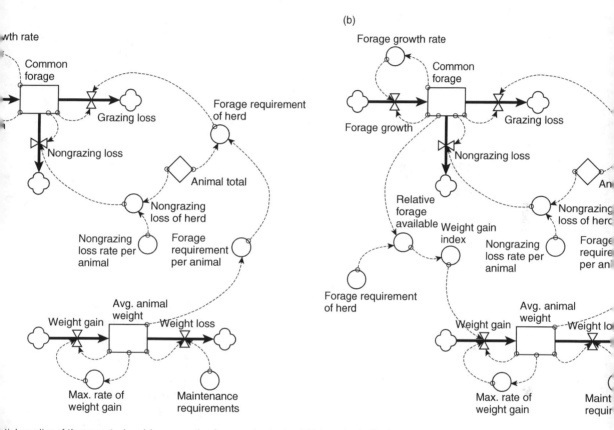

wth rate

Common
forage

Grazing loss

Nongrazing loss

Forage requirement
of herd

Animal total

Nongrazing
loss of herd

Nongrazing
loss rate per
animal

Forage
requirement
per animal

Avg. animal
weight

Weight gain   Weight loss

Max. rate of
weight gain

Maintenance
requirements

(b)

Forage growth rate

Common
forage

Forage growth

Grazing loss

Nongrazing loss

An

Relative
forage
available   Weight gain
index

Nongrazing
loss of herc

Nongrazing
loss rate per
animal

Forage
require
per an

Forage requirement
of herd

Avg. animal
weight

Weight gain   Weight lo

Max. rate of
weight gain

Maint
requir

tial coupling of the conceptual models representing forage and animal weight dynamics on the Commons.

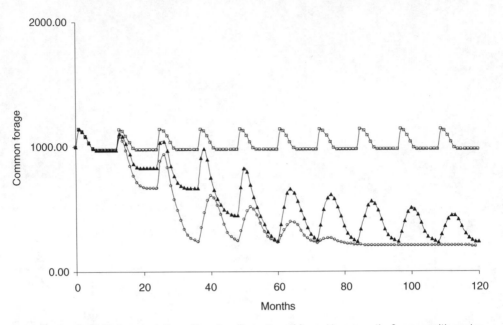

Fig. 2.11 Our expectations of long-term fluctuations of forage biomass on the Commons with number of grazing animals constant at two (open squares), increasing by one each year (closed triangles), and increasing by two each year (open circles).

month should provide adequate detail to temporal patterns in these components for our purposes (II_b)(IDM_1).

Now we are ready to begin our calculations. But since we are dealing with even more calculations than in the population example, we decide not to try to base the first projection on all of them. Almost surely we will make some errors in our initial calculations, and these will be almost impossible to track down among so many interdependent calculations. Rather, as in the population example, we decide to begin with some very simple calculations, and then proceed step by step, adding a bit more complexity at each step. First, we will focus on calculations related to forage dynamics, then on calculations related to animal weight dynamics, and then on coupling these two sets of calculations. As in the population example, we will check for potential errors at each step by comparing the results of our calculations to our expectations. And once again, depending on the simplifying assumptions we have made for any given set of calculations, we may alter our expectations from those we described for the complete set of calculations.

We begin with calculations related to Fig. 2.8b. We will compare the results of these calculations to results of the "no grazing" treatment of the experiment for which we have data (Fig. 2.7, open squares). Based on Fig. 2.8b, we know we can calculate changes in forage biomass by adding sequential increments of forage growth. We know that forage growth rate decreases linearly from a maximum of 1 unit of biomass produced

COMMON-SENSE SOLUTIONS: THREE EXERCISES

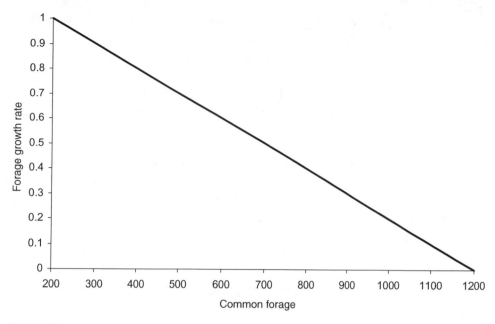

Fig. 2.12 Relationship between forage growth rate (units of biomass growth/unit of forage biomass–month) and forage (units of biomass). The line illustrates the source of values for the calculations in Table 2.2.

per unit of forage biomass per month to 0 as forage biomass increases from 200 to 1200 units (Fig. 2.12) (II$_c$, II$_d$)(*IDM*$_1$). We decide to initialize forage at 100 biomass units, as in the experiment (II$_e$)(*IDM*$_1$). Our first calculation of monthly forage growth is:

> Forage growth during 1st month
> =(1 unit of growth/unit of forage) ∗ (100 units of forage)
> =100 units of growth   (eq. 2.22)

We calculate the amount of forage at the beginning of the second month by adding the growth during the first month to the amount of biomass that was already in the Commons (at the beginning of the first month) (eq. 2.23).

> Forage biomass at beginning of 2nd month
> =(100 units of forage)+(100 units of forage)
> =200 units of forage   (eq. 2.23)

Likewise, we calculate the amount of forage at the beginning of the third month by calculating the amount of growth during the second month (eq. 2.24), and then adding it to the forage that was present at the beginning of the second month (eq. 2.25).

Forage growth during 2nd month
$$= (1 \text{ unit of growth/unit of forage}) \star (200 \text{ units of forage})$$
$$= 200 \text{ units of growth} \qquad \text{(eq. 2.24)}$$

Forage biomass at beginning of 3rd month
$$= (200 \text{ units of forage}) + (200 \text{ units of growth})$$
$$= 400 \text{ units of forage} \qquad \text{(eq. 2.25)}$$

Continuing in the same manner, we calculate the amount of forage at the beginning of the fourth month by calculating the amount of growth during the third month (eq. 2.26), and then adding it to the forage that was present at the beginning of the third month (eq. 2.27). Note that beginning in month 3, monthly forage growth rate begins to decline because forage biomass has become greater than 200 units (Fig. 2.12).

Forage growth during 3rd month
$$= (0.80 \text{ units of growth/unit of forage}) \star (400 \text{ units of forage})$$
$$= 320 \text{ units of growth} \qquad \text{(eq. 2.26)}$$

Forage biomass at beginning of 4th month
$$= (400 \text{ units of forage}) + (320 \text{ units of growth})$$
$$= 720 \text{ units of forage} \qquad \text{(eq. 2.27)}$$

We repeat these calculations in the same manner for months 4 through 12. Results from these subsequent calculations are presented in Table 2.2.

We now compare the results of our calculations to results of the "no grazing" treatment of the 1-year grazing experiment. Our calculations

Table 2.2 Results of calculations of forage dynamics related to Fig. 2.8b.

Month	Forage growth rate	Forage growth	Forage biomass
1	1	100	100
2	1	200	200
3	0.8	320	400
4	0.48	345.6	720
5	0.13	143.22	1065.60
6	0	0	1208.82
7	0	0	1208.82
8	0	0	1208.82
9	0	0	1208.82
10	0	0	1208.82
11	0	0	1208.82
12	0	0	1208.82
Final			1208.82

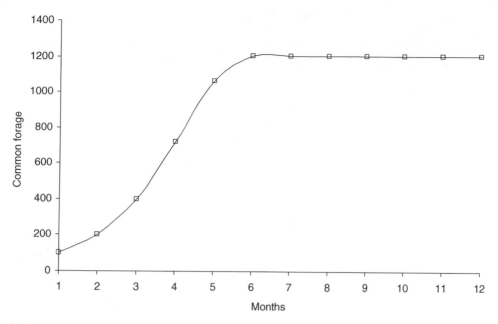

Fig. 2.13 Comparison of our calculations (solid line) to observed results (open squares) from the 1-year grazing experiment with no grazing animals present. (The two lines overlap exactly.)

seem reasonable $(III_a)(IDM_1)$, and our results correspond well both qualitatively $(III_b)(IDM_1)$ and quantitatively $(III_c)(IDM_1)$ with our expectations (Fig. 2.13). So, we will proceed.

Now, according to our plan, we should continue with calculations related to forage dynamics, that is, we should add either grazing loss or nongrazing loss to our calculations. However, upon reflection, we decide it might be more convenient to change the plan and proceed with the calculations related to animal weight dynamics, and then return to grazing and nongrazing losses.

Thus we proceed with calculations related to Fig. 2.9b. We will compare the results of these calculations to our expectations for animal growth (continuous linear weight gain from 10 to 90 biomass units). However, since these calculations do not include weight losses, and assume no shortage of forage, they should yield weights somewhat higher than those we expect when these losses are included. Nonetheless, the comparison should provide a useful checkpoint. Based on Fig. 2.9b, we know we can calculate changes in animal weight by adding sequential increments of weight gain. We know the maximum rate of monthly weight gain decreases linearly from 1.25 times current body weight to 0 as weight increases from 5 (the lower lethal body weight) to 100 units (Fig. 2.14) $(II_c, II_d)(IDM_2)$. We decide to initialize weight at 10 biomass units, the weight at which we put animals on the Commons $(II_e)(IDM_2)$. Our first calculation of monthly weight gain is 11.84 units (eq. 2.28):

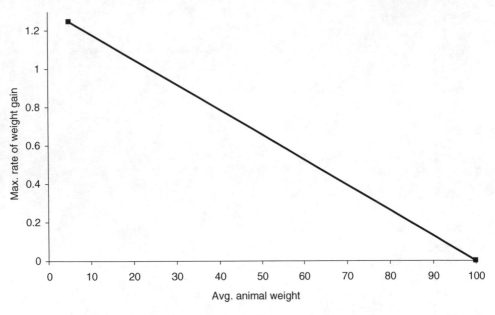

Fig. 2.14 Relationship between the maximum rate of weight gain (units of weight gain/unit of animal weight-month) and current animal weight (units of weight). Specific points on the line illustrate the source of values for the calculations in Table 2.3.

Weight gain during 1st month
= (1.184 units of weight gain/unit of animal weight)
⋆ (10 units of animal weight)
= 11.84 units of weight gain   (eq. 2.28)

We calculate animal weight at the beginning of the second month by adding the weight gain during the 1st month to the animal weight at the beginning of the first month (eq. 2.29).

Animal weight at beginning of 2nd month
= (10 units of weight) + (11.84 units of weight gain)
= 21.84 units of animal weight   (eq. 2.29)

We repeat these calculations in the same manner for the third month.

Weight gain during 2nd month
= (1.03 units of weight gain/unit of animal weight)
⋆ (21.84 units of animal weight)
= 22.46 units of weight gain   (eq. 2.30)

and add it to the weight at the beginning of the second month (eq. 2.31).

Animal weight at beginning of 3rd month

$$= (21.84 \text{ units of weight}) + (22.46 \text{ units of weight gain})$$
$$= 44.3 \text{ units of animal weight} \qquad \text{(eq. 2.31)}$$

We repeat these calculations in the same manner for months 3 through 12. Results from these subsequent calculations are presented in Table 2.3.

We now compare results of these calculations to our expectations for animal growth. Our calculations seem reasonable (III_a)(IDM_2). However, our calculated growth curve looks different, both qualitatively (III_b)(IDM_2) and quantitatively (III_c)(IDM_2), from our expectation (Fig. 2.15). Thus, we are obliged to reconcile this difference before we proceed.

We recognize that, in general, there are three possible explanations for differences between our expectations and our calculations:

1 Our calculations are flawed, either conceptually or numerically.
2 Our expectation is flawed, either conceptually or numerically.
3 Both are somehow flawed.

In this case, after reexamining the relationship between the maximum rate of weight gain and animal weight, and recalling our animals always grow proportionally faster when they are small, we decide our expectation of linear growth was flawed. We should have expected that weight would increase in an S-shaped manner, as indicated by our calculations. Thus we alter our expectation and proceed.

We proceed with calculations related to weight losses (Fig. 2.9c). We will compare the results of these calculations to our new expectations for animal growth. Since these calculations will include weight losses, although we still will assume no shortage of forage, we expect they will

Table 2.3 Results of calculations of weight dynamics related to Fig. 2.9b.

Month	Max. rate of weight gain	Weight gain	Weight
1	1.18	11.84	10
2	1.03	22.46	21.84
3	0.73	32.47	44.3
4	0.31	23.46	76.77
5	0	0	100.24
6	0	0	100.24
7	0	0	100.24
8	0	0	100.24
9	0	0	100.24
10	0	0	100.24
11	0	0	100.24
12	0	0	100.24
Final			100.24

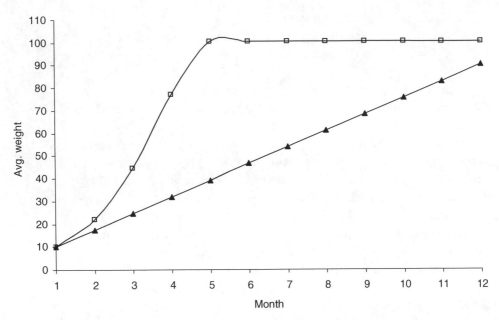

Fig. 2.15 Comparison of our calculations of animal growth (without including weight loss due to maintenance requirements, and assuming no shortage of forage) (open squares) to our expectation of the manner in which the animals should grow (closed triangles).

yield weights somewhat lower than the S-shaped curve in Fig. 2.15. Based on Fig. 2.9c, we know we now need to calculate changes in animal weight by subtracting weight losses, as well as adding weight gains. We know the animals normally metabolize 10% of their weight each month due to energy expenditures to meet maintenance requirements. We also know that energy expenditures of females increase by 50% during the 6 months they are pregnant, and by 100% during the 3 months they are nursing their two young. However, since individuals become sexually mature at 18–20 months of age, and we always harvest our animals at 14 months of age, this information regarding additional energy expenditures for reproduction now seems irrelevant. We decide to calculate maintenance requirements simply as 10% of body weight per month (II_c, II_d)(IDM_3).

We again initialize animal weight at 10 biomass units (II_e)(IDM_3), and calculate weight gain for the first month (eq. 2.32)

> Weight gain during 1st month
> = (1.184 units of weight gain/unit of animal weight)
> ⋆ (10 units of animal weight)
> = 11.84 units of weight gain   (eq. 2.32)

In order to determine the weight at beginning of the second month, we need to also calculate the weight loss for the first month (eq. 2.33)

Weight loss during 1st month
   =(0.1 ★ units of weight loss/ unit of animal weight)
    ★ (10 units of animal weight)
   =1 unit of weight loss   (eq. 2.33)

We calculate animal weight at the beginning of the second month by adding the weight gain during the first month, and subtracting the weight loss during the first month, to the animal weight at the beginning of the first month (eq. 2.34).

Animal weight at beginning of 2nd month
   =(10 units of animal weight)+[(11.84 units of weight gain)
    −(1 unit of weight loss)]
   =20.84 units of animal weight   (eq. 2.34)

We repeat these calculations in the same manner for months 2 through 12. Results from these subsequent calculations are presented in Table 2.4.

We now compare results of these calculations to our new expectations for animal growth. Our calculations seem reasonable (III$_a$)(IDM_3), the growth curve based on calculations including weight loss is somewhat lower than the curve based on calculations without weight loss included (III$_c$)(IDM_3), and the curve retains its sigmoid shape (Fig. 2.16) (III$_b$)(IDM_3). So we will proceed.

Both forage dynamics and animal weight dynamics are behaving reasonably in an uncoupled fashion, that is, assuming the animals are not affecting forage dynamics (Fig. 2.8c) and forage dynamics are not affecting animal weight dynamics (Fig. 2.9c). We will proceed to couple the two sets of calculations by first representing the effect of grazing and tram-

Table 2.4 Results of calculations of weight dynamics related to Fig. 2.9c.

Month	Max. rate of weight gain	Weight gain	Rate of weight loss	Weight loss	Weight
1	1.18	11.84	0.1	1	10
2	1.04	21.71	0.1	2.08	20.84
3	0.78	31.7	0.1	4.05	40.47
4	0.42	28.58	0.1	6.81	68.12
5	0.13	11.97	0.1	8.99	89.88
6	0.09	8.72	0.1	9.29	92.86
7	0.1	9.35	0.1	9.23	92.3
8	0.1	9.22	0.1	9.24	92.42
9	0.1	9.25	0.1	9.24	92.4
10	0.1	9.24	0.1	9.24	92.4
11	0.1	9.24	0.1	9.24	92.4
12	0.1	9.24	0.1	9.24	92.4
Final					92.4

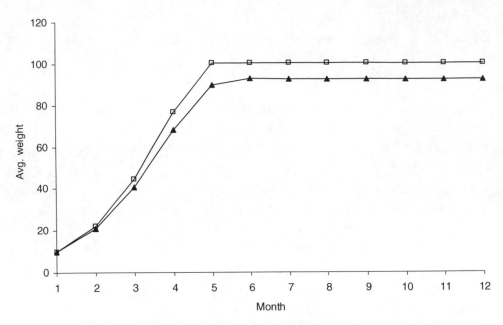

Fig. 2.16 Comparison of our calculations of animal growth with (closed triangles) and without (open squares) including weight loss due to maintenance requirements.

pling (nongrazing loss) on forage dynamics (Fig 2.10a), still assuming forage dynamics are not affecting animal weight dynamics.

We will compare the results of these calculations to results of the two "grazing" treatments of the experiment for which we have data (Fig. 2.7, closed triangles, open circles). We expect that our calculated forage curves should be similar to the experimental curves, or perhaps somewhat lower. If growth of the experimental animals was limited by available forage, they would have been smaller and, hence, eaten less, than our "calculated" animals. Based on Fig. 2.10a, we know we now need to calculate changes in forage biomass by subtracting forage losses due to trampling and grazing, as well as adding forage growth. We also need to represent more than 1 animal on the Commons.

We have guessed that each animal tramples 2% of the forage each month, and we know that each animal requires (wants to eat) 1 unit of forage biomass per unit of animal biomass per month. We will assume that trampled forage can not be eaten, thus we will calculate grazing and trampling losses as completely additive (II_c, II_d)(IDM_4). We also will calculate grazing losses assuming that animals are growing at their maximum rates (since we still are assuming forage dynamics are not affecting animal weight dynamics). We again initialize forage biomass at 100 units, as in the 1-year grazing experiment (II_e)(IDM_4). Our first calculation of monthly forage growth (eq. 2.35) is exactly as before (eq. 2.22):

Forage growth during 1st month
    =(1 unit of growth/unit of forage)*(100 units of forage)
    =100 units of growth   (eq. 2.35)

Using our guess that each animal tramples 2% of the forage each month, we next calculate the forage loss due to trampling (eq. 2.36).

Forage trampling loss during 1st month
    =(0.02 of forage lost to trampling/animal)
    (1 animal)(100 units of forage)
    =2 units of forage loss to trampling   (eq. 2.36)

We then calculate the forage loss due to grazing during the first month (eq. 2.37):

Forage grazing loss during 1st month
    =(1 unit of forage loss to grazing/unit of animal biomass)
    (10 units of animal biomass/animal)(1 animal)
    =10 units of forage loss to grazing   (eq. 2.37)

We then calculate forage biomass at the beginning of the second month by adding the growth to, and subtracting the grazing and trampling losses from, the forage on the Commons at the beginning of the first month (eq. 2.38).

Forage biomass at beginning of 2nd month
    =(100 units of forage)+[(100 units of forage growth)
    −(2 units of forage loss to trampling)
    −(10 units of forage loss to grazing]
    =188 units of forage biomass   (eq. 2.38)

We repeat these calculations in the same manner for months 2 through 12, and for months 1 through 12 with 2 grazing animals. Results from these subsequent calculations are presented in Table 2.5.

We now compare results of these calculations to results of the two "grazing" treatments of the 1-year grazing experiment. Our calculations seem reasonable (III_a)(IDM_4), and we note that calculated and observed forage curves are quite similar for both grazing treatments (III_b)(IDM_4) (Fig. 2.17). However, our curve calculated with 2 animals is a bit lower than observed during months 4 through 8 (III_c)(IDM_4), suggesting that growth of the experimental animals, indeed, might have been limited by available forage during this period (Fig. 2.17b). We decide to proceed.

We now add the effect of forage dynamics on animal weight dynamics (Fig 2.10b). We again will compare the results of these calculations to results of the two "grazing" treatments of the 1-year grazing experiment (Fig. 2.7, closed triangles, open circles). Our expectation is that our curve

Table 2.5 Results of calculations of forage dynamics related to Fig. 2.17a with (a) one grazing animal and (b) two grazing animals.

Table 2.5a

Month	Forage growth rate	Forage growth	Forage requirement per animal	Grazing loss	Nongrazing loss per animal	Nongrazing loss	Forage biomass
1	1	100	10	10	0.02	2	100
2	1	188	20.84	20.84	0.02	3.76	188
3	0.85	298.2	40.47	40.47	0.02	7.03	351.4
4	0.6	360	68.12	68.12	0.02	12.04	602.1
5	0.32	280.51	89.88	89.88	0.02	17.64	881.94
6	0.15	153.04	92.86	92.86	0.02	21.1	1055
7	0.11	115.95	92.3	92.3	0.02	21.88	1094
8	0.1	114.2	92.42	92.42	0.02	21.92	1096
9	0.1	114.34	92.4	92.4	0.02	21.91	1096
10	0.1	114.31	92.4	92.4	0.02	21.91	1096
11	0.1	114.31	92.4	92.4	0.02	21.91	1096
12	0.1	114.31	92.4	92.4	0.02	21.91	1096
Final							1096

Table 2.5b

Month	Forage growth rate	Forage growth	Forage requirement per animal	Grazing loss	Nongrazing loss per animal	Nongrazing loss	Forage biomass
1	1	100	10	20	0.02	4	100
2	1	176	20.84	41.68	0.02	7.04	176
3	0.9	271.95	40.47	80.93	0.02	12.13	303.28
4	0.72	346.12	68.12	136.24	0.02	19.29	482.17
5	0.53	354.71	89.88	179.76	0.02	26.91	672.76
6	0.38	311.25	92.86	185.72	0.02	32.83	820.79
7	0.29	261.72	92.3	184.6	0.02	36.54	913.49
8	0.25	234.63	92.42	184.84	0.02	38.16	954.08
9	0.23	226.26	92.4	184.79	0.02	38.63	965.7
10	0.23	224.17	92.4	184.8	0.02	38.74	968.54
11	0.23	223.71	92.4	184.8	0.02	38.77	969.18
12	0.23	223.60	92.4	184.80	0.02	38.77	969.32
Final							969.32

calculated with 2 animals now will be even more similar to the observed curve. We also will compare our calculations to our expectation of long-term forage dynamics based on our personal observations of the Commons (Fig. 2.11), and to our expectation that animals will weigh about 90 biomass units after 1 year on the Commons. Based on Fig. 2.10b, we know we now need to calculate actual animal weight gain depending on the relative availability of forage. We know as the ratio of forage biomass available to forage biomass required decreases from 5 to 1, the proportion of maximum possible weight gain realized decreases from 1

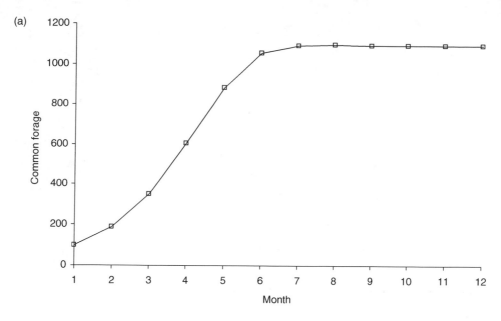

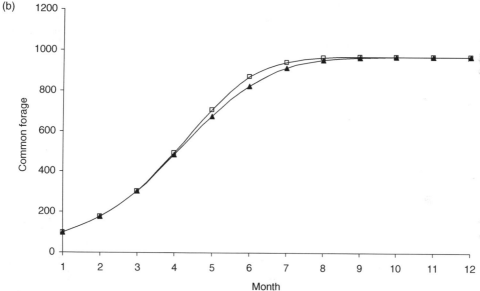

Fig. 2.17 Comparison of our calculations of forage dynamics (open squares) to forage dynamics observed (closed triangles) during the 1-year grazing experiment with (a) one grazing animal (the two lines overlap exactly) and (b) two grazing animals. Our calculations assume that grazing animals gained weight at their maximum rates throughout the year.

to 0 (Fig. 2.18) (II_c, II_d)(IDM_{last}). Thus, we will initialize animal weight at 10 biomass units, forage biomass at 100 units, and perform our calculations of forage growth, forage loss due to trampling, forage loss due to grazing, and animal weight loss as before (II_e)(IDM_{last}). Weight gain will now be calculated by taking into account the proportion of maximum possible weight gain attained (eq. 2.39).

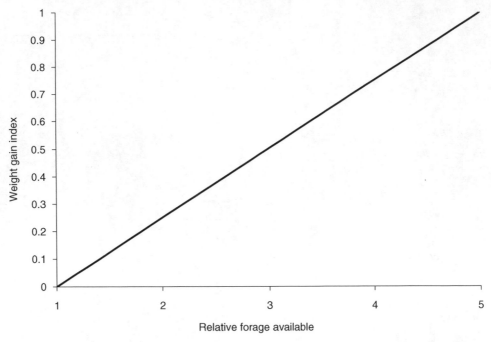

Fig. 2.18 Relationship between the proportion of maximum possible weight gain actually attained and the relative availability of forage (forage biomass available/forage biomass required to meet animal requirements).

> Weight gain during 1st month
> $= (1.184$ units of weight gain/unit of animal weight$)$
> $\star (10$ units of animal weight$) \star (1.0$: proportion of maximum possible weight gain attained$)$
> $= 11.84$ units of weight gain (eq. 2.39)

We then calculate the values of forage biomass and animal weight at the beginning of the second month as before. Results of these calculations for 1 grazing animal, and for 2 grazing animals, are presented in Table 2.6. Note that during months 2 through 6 of the calculations for 2 animals, the proportion of maximum possible weight gain actually attained is less than 1.

We now compare results of these calculations to results of the two "grazing" treatments of the 1-year grazing experiment. Our calculations seem reasonable $(III_a)(IDM_{last})$, and now both of the calculated forage curves are quite similar to the curves based on the experiment $(III_b, III_c)(IDM_{last})$ (Fig. 2.19). Final animal weight also is quite similar to our expectation (Table 2.6).

We next compare our long-term calculations to our expectations of long-term (10-year) forage and animal weight dynamics. We perform these long-term calculations exactly as we did the 12-month calculations,

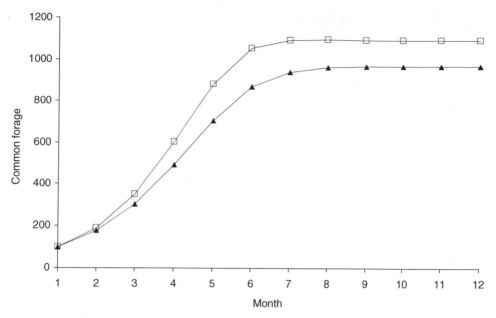

Fig. 2.19 Comparison of our calculations of forage dynamics to forage dynamics observed during the 1-year grazing experiment with one grazing animal (solid line and open squares; the two lines overlap exactly) and two grazing animals (solid line and closed triangles; the two lines overlap exactly). Our calculations now include the effect of shortage of available forage on animal weight dynamics.

except we initialize forage biomass at 1000 units, to represent current conditions on the Commons, rather than at 100 units, which represented the grazing experiments. We also will reinitialize individual animal weight at 10 biomass units each January (once every 12 sets of calculations), to represent the replacement of new animals on the Commons (II_e)(IDM_{last}). These calculations seem reasonable (III_a)(IDM_{last}), and results also are similar to our expectations in that forage biomass fluctuates seasonally close to 1000 biomass units and animals weigh about 90 biomass units at harvest (Fig. 2.20) (III_b, III_c)(IDM_{last}). We decide our calculations provide a useful representation of the dynamics of the Commons. That is, they represent the components we are interested in projecting into the future, and the most important processes that make them change over time.

We now are almost ready to make our projections into the future under different scenarios. However, since the scenarios of interest will represent the addition of quite a few more animals to the Commons, we are a bit concerned about having to guess at trampling rate. We decide to see how sensitive our long-term calculations of forage biomass and animal weight dynamics are to different estimates of trampling rate with different numbers of animals on the Commons (III_d)(IDM_{last}). Our best guess is that each animal tramples 2% of the current forage per month. We decide to repeat our calculations assuming a trampling rate of 5% per month. Based on what we do know about the animals and the forage, it

Table 2.6 Results of calculations of forage and animal weight dynamics related to Fig. 2.19 with (a) 1 grazing animal and (b) 2 grazing animals.

Table 2.6a

Month	Forage growth rate	Forage growth	Forage requirement per animal	Grazing loss	Nongrazing loss per animal	Nongrazing loss	Forage biomass	Animal weight	Weight gain index
1	1	100	10	10	0.02	2	100	10	1
2	1	188	20.84	20.84	0.02	3.76	188	20.84	1
3	0.85	298.2	40.47	40.47	0.02	7.03	351.4	40.47	1
4	0.6	360	68.12	68.12	0.02	12.04	602.1	68.12	1
5	0.32	280.51	89.88	89.88	0.02	17.64	881.94	89.88	1
6	0.15	153.04	92.86	92.86	0.02	21.1	1054.93	92.86	1
7	0.11	115.95	92.3	92.3	0.02	21.88	1094.01	92.3	1
8	0.1	114.2	92.42	92.42	0.02	21.92	1095.79	92.42	1
9	0.1	114.34	92.4	92.4	0.02	21.91	1095.64	92.4	1
10	0.1	114.31	92.4	92.4	0.02	21.91	1095.67	92.4	1
11	0.1	114.31	92.4	92.4	0.02	21.91	1095.67	92.4	1
12	0.1	114.31	92.4	92.4	0.02	21.91	1095.67	92.4	1
Final							1095.67	92.4	

Table 2.6b

Month	Forage growth rate	Forage growth	Forage requirement per animal	Grazing loss	Nongrazing loss per animal	Nongrazing loss	Forage biomass	Animal weight	Weight gain index
1	1	100	10	20	0.02	4	100	10	1
2	1	176	20.84	41.68	0.02	7.04	176	20.84	0.81
3	0.9	271.95	36.24	72.49	0.02	12.13	303.28	36.24	0.8
4	0.71	348.03	56.82	113.64	0.02	19.62	490.61	56.82	0.83
5	0.49	348.9	77.91	155.82	0.02	28.22	705.38	77.91	0.88
6	0.33	286.97	90.09	180.17	0.02	34.81	870.24	90.09	0.96
7	0.26	242.88	92.33	184.66	0.02	37.69	942.23	92.33	1
8	0.24	228.41	92.42	184.83	0.02	38.51	962.76	92.42	1
9	0.23	224.71	92.4	184.79	0.02	38.71	967.82	92.4	1
10	0.23	223.82	92.4	184.8	0.02	38.76	969.02	92.4	1
11	0.23	223.63	92.4	184.8	0.02	38.77	969.28	92.4	1
12	0.23	223.63	92.4	184.8	0.02	38.77	969.34	92.4	1
Final							969.34	92.4	

seems quite unlikely that trampling rate could be more than this. We also will repeat our calculations assuming a trampling rate of 0, which, obviously, is the lower limit. Finally, in addition to repeating our calculations with 5% and 0% trampling rates, assuming 2 animals are on the Commons, we will repeat them for 5%, 2%, and 0% trampling rates, assuming 10 animals are on the Commons. In the scenario in which we keep pace with our neighbors, there will be 10 animals on the Commons after 5 years. Results of these calculations indicate, as we suspected, that forage biomass and animal weight dynamics are relatively insensitive to changes

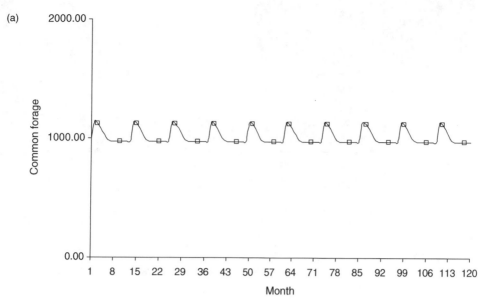

(a)

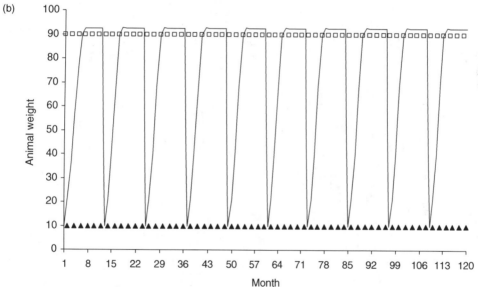

(b)

Fig. 2.20 (a) Comparison of our long-term (10-year) calculations of forage dynamics (solid line) to our expectations (open squares). (b) Comparison of long-term calculations of animal weight dynamics on the Commons (solid line) to expected starting (solid triangles) and ending weights (open squares).

in our estimate of trampling rates with only 2 animals on the Commons (Fig. 2.21). However, the dynamics of both forage biomass and animal weight are very sensitive to changes in our estimate of trampling rates with 10 animals on the Commons (Fig. 2.22). Based on these results we conclude that we should give high priority to gathering information that would allow us to estimate trampling rate with more confidence. We also conclude that, in the meantime, it would be wise to make alternative

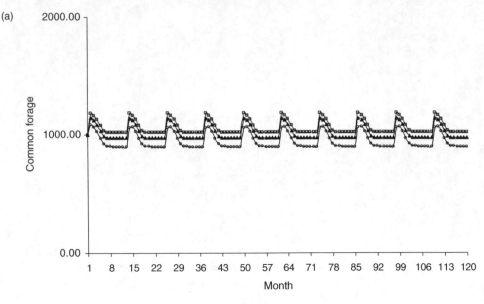

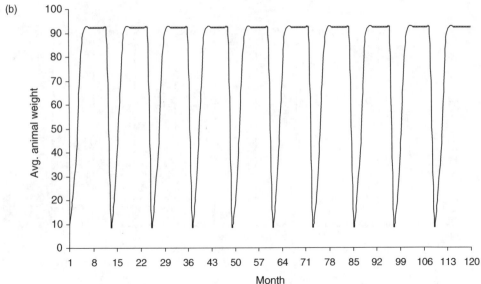

Fig. 2.21 Sensitivity of our long-term (10-year) calculations of (a) forage dynamics and (b) animal weight dynamics with two animals on the Commons to changes in our estimation of forage trampling rate (all three lines overlap in b). Open squares, closed triangles, and open circles represent 0%, 2%, and 5%, respectively, of forage biomass trampled per animal per month.

projections of each of our scenarios using a range of trampling rates that appropriately reflects the degree of our uncertainty in its true value. We decide to use 0%, 2%, and 5% per animal per month.

Project scenarios We now are ready to make our projections into the future under different scenarios to see (1) if forage biomass will decrease over the next 10 years, and (2) if the final weights of our animals will decrease

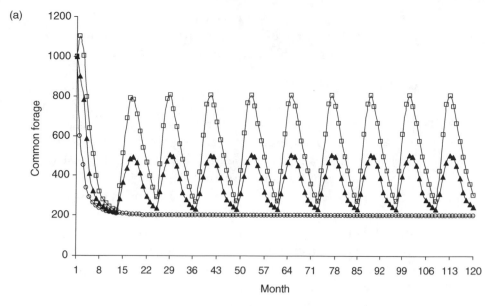

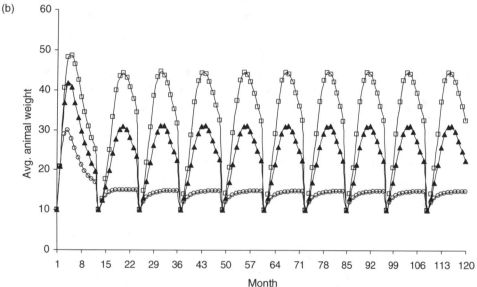

Fig. 2.22 Sensitivity of our long-term (10-year) calculations of (a) forage dynamics and (b) animal weight dynamics with 10 animals on the Commons to changes in our estimation of forage trampling rate. Open squares, closed triangles, and open circles represent 0%, 2%, and 5%, respectively, of forage biomass trampled per animal per month.

over the next 10 years. We recall the two scenarios of immediate interest are (1) if we increase the number of animals we put on the Commons at the same rate our neighbor does (add 1 additional animal each year), and (2) if we continue to put just 1 animal on the Commons each year $(IV_a)(FM)$. We perform the calculations for these projections in exactly the same manner that we performed our previous long-term calculations, except we put either 1 or 2 additional animals on the Commons each

January (once every 12 sets of calculations), depending on the scenario we are projecting. And, of course, we repeat the calculations for each scenario using each of the three estimates of trampling rate.

Answer Results of our calculations suggest that if our neighbor adds an additional animal to the Commons each year, regardless of whether we add an additional animal each year, neither forage nor animal production can be sustained. These results also suggest that both forage biomass and individual animal weight at harvest decrease under each of the two scenarios, regardless of our estimate of trampling rate (Figs 2.23, 2.24). Thus, we have confidence in these general trends in spite of our uncertainty about trampling rate $(IV_b)(FM)$. These results also seem reasonable when compared to our initial expectations of forage dynamics with an increasing number of animals on the Commons (Figs 2.11, 2.23a, 2.24a), and the logic of our calculations still appears to be sound. Our calculated animal weights at harvest after 10 years of increasing by 2 (Fig. 2.23b), or 1 (Fig. 2.24b), the number of animals on the Commons are noticeably lower than our initial guesses (20 and 45 weight units, respectively). However, since we really did not have much confidence in our initial guesses, we decide our calculated harvest weights probably provide better estimates.

Results of our calculations also suggest that we might be able to refine our estimate of trampling rate rather quickly if we monitor harvest weights of animals in the real system carefully, since harvest weights calculated using the different trampling rates begin to diverge noticeably after just a couple of years. Forage biomass on the Commons calculated using the different trampling rates also begins to diverge noticeably after just a couple of years, but monitoring harvest weights is a lot easier than sampling forage biomass.

Further questions/answers At this point, we have answered, at least tentatively, our initial questions. But, as often happens, as we have answered our initial questions, we have raised some others. First, since individual animal harvest weights do not decline drastically during the first year or two, and our neighbor seems so intensely focused on increasing the total biomass of his herd, we wonder if the additional biomass resulting from having more animals in the harvest will prevent him from noticing the decreasing individual weights. We can calculate total animal biomass by simply multiplying individual animal weight each month by the number of animals on the Commons. We are somewhat discouraged to find that, although total animal biomass harvested decreases in the long term, total harvest increases for 1 year under all scenarios and trampling rates, and for 2 or 3 years under some combinations of scenario and trampling rate (Fig. 2.25). This may erroneously lead our neighbor into thinking his plan is working quite well, at least during the early years.

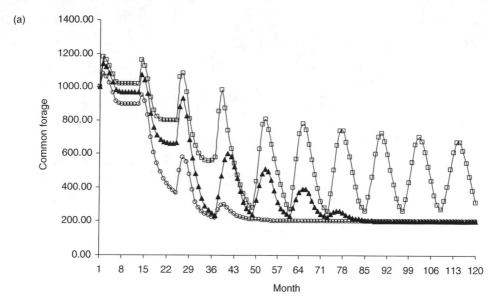

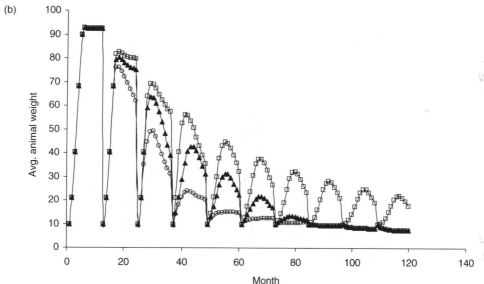

Fig. 2.23 Results of calculations of (a) forage biomass and (b) animal weight dynamics over the next 10 years, assuming we increase the number of animals we put on the Commons at the same rate as our neighbor does (add one additional animal each year). Open squares, closed triangles, and open circles represent 0%, 2%, and 5%, respectively, of forage biomass trampled per animal per month.

Another trend in our initial results that catches our eye is the seasonality of individual animal weight dynamics. We notice that animals reach their maximum weight before harvest and then actually lose weight. The weight loss is not very noticeable with only 2 animals on the Commons (Figs 2.20b, 2.21b), but becomes increasingly noticeable as the number of animals increases (Figs 2.22b, 2.23b, 2.24b). This suggests some additional scenarios that would be interesting to explore. We decide to calculate

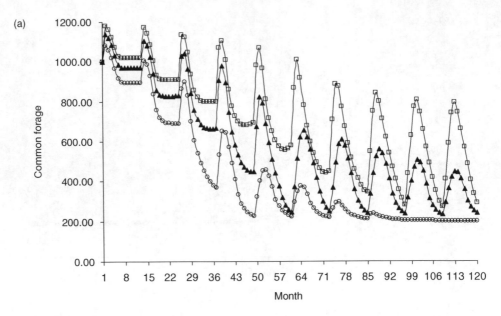

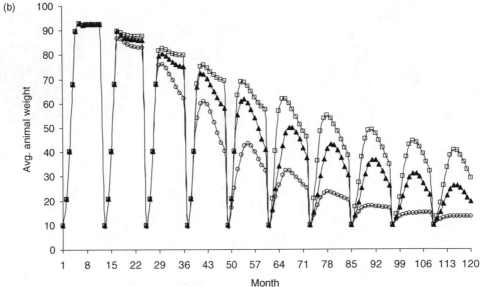

Fig. 2.24 Results of calculations of (a) forage biomass and (b) animal weight dynamics over the next 10 years, assuming we continue to put just one animal on the Commons each year, while our neighbor adds one additional animal each year. Open squares, closed triangles, and open circles represent 0%, 2%, and 5%, respectively, of forage biomass trampled per animal per month.

what would happen if we added the same number of animals as our neighbor, but we both harvested our animals earlier in the year, say, in June. We perform these calculations in exactly the same manner as before, except we reset individual animal weight to 0 each July (and, as before, to 10 in January). Our calculations indicate that forage biomass can recover completely each year with the 6-month no-grazing period, except under the

COMMON-SENSE SOLUTIONS: THREE EXERCISES

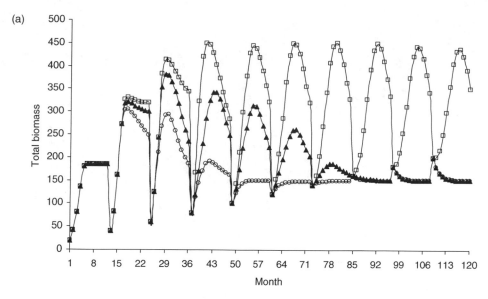

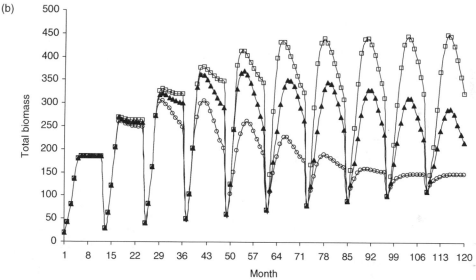

Fig. 2.25 Estimates of total animal biomass over the next 10 years, assuming (a) we increase the number of animals we put on the Commons at the same rate as our neighbor does (add one additional animal each year), and (b) we continue to put just one animal on the Commons each year, while our neighbor adds one additional animal each year. Open squares, closed triangles, and open circles represent 0%, 2%, and 5%, respectively, of forage biomass trampled per animal per month.

highest trampling rate estimate (Fig. 2.26a). Total animal biomass harvested also can be sustained over the 10-year period, except under the highest trampling rate estimate (Fig. 2.27). However, harvest weights still decrease continually, regardless of our estimate of trampling rate (Fig. 2.26b). We also notice that harvest weight in year 10 of the calculations assuming 5% trampling rate is 5.9 biomass units, which is becoming dangerously close to the lower lethal body weight (5 biomass units).

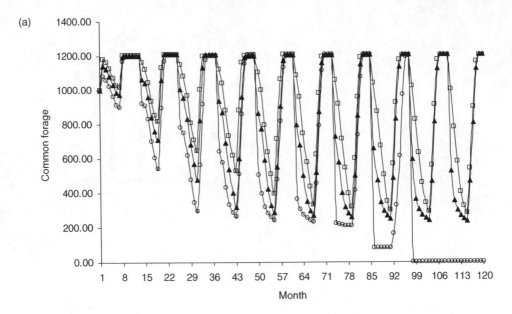

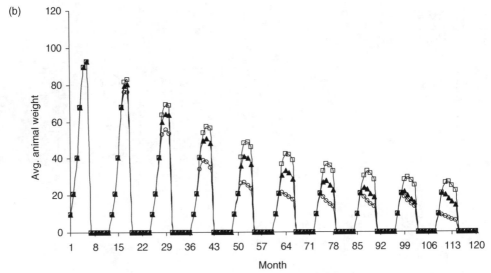

Fig. 2.26 Results of calculations of (a) forage biomass and (b) animal weight dynamics over the next 10 years, assuming we increase the number of animals we put on the Commons at the same rate as our neighbor does (add one additional animal each year), but that we both harvest our animals in June rather than December. Open squares, closed triangles, and open circles represent 0%, 2%, and 5%, respectively, of forage biomass trampled per animal per month.

Armed with these results, and perhaps others that occur to us, and most importantly with our set of calculations with which we can project other animal stocking scenarios, perhaps we can suggest to our neighbor that he should reconsider his plan $(IV_c)(FM)$.

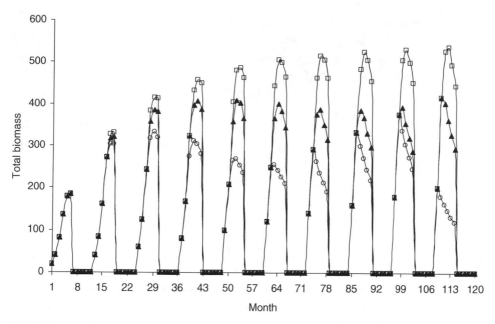

Fig. 2.27 Estimates of total animal biomass over the next 10 years, assuming we increase the number of animals we put on the Commons at the same rate as our neighbor does (add one additional animal each year), but that we both harvest our animals in June rather than December. Open squares, closed triangles, and open circles represent 0%, 2%, and 5%, respectively, of forage biomass trampled per animal per month.

The systems approach is both a philosophical perspective and a collection of techniques, including simulation, developed explicitly to address problems dealing with complex systems. The systems approach emphasizes a holistic approach to problem solving as opposed to the more traditional reductionist approach. The use of mathematical models to identify and simulate important characteristics of complex systems is an important component of the systems approach. In the simplest sense, a system is any set of objects that interact. A mathematical model is a set of equations that describes the interrelationships among these objects. By solving the equations comprising a mathematical model, we can mimic, or simulate, the dynamic (time-varying) behavior of the system.

The basic approach is to (1) develop a conceptual model (generally a box and arrow diagram) identifying specific cause–effect relationships among important components of the system in which we are interested, (2) quantify (write mathematical equations for) these relationships based on the analysis of the best information available, (3) evaluate the usefulness of the model ("validate" model) in terms of its ability to simulate known system behavior, and (4) apply the model (conduct simulated experiments) to address our questions concerning unknown system behavior (Fig. 2.1, center portion). The idea is that we conduct experiments with the model exactly as we would conduct experiments in the

2.2 The systems approach to problem solving

field or laboratory (Fig. 2.28, modified from Van Dyne, 1969). Just as we "abstract" particular parts of the world to permit more detailed studies in the field or laboratory, we abstract important real-system components and processes in the form of a simulation model to allow more controlled observation and experimentation. We develop an experimental design for simulations in exactly the same manner that we do for field experiments. Likewise, we analyze simulation results using the same qualitative, quantitative, and/or statistical methods that we use to analyze results from field experiments.

2.2.1 The conceptual model (Phase I)

In the first phase, we develop a conceptual, or qualitative, model of the system-of-interest (Fig. 2.1). Based on the objectives of the modeling project, we decide which components in the real-world system should be included in our system-of-interest and how they should be related to one another. We represent these components and their relationships, which collectively form our conceptual model, diagrammatically using symbols that indicate the specific nature of the relationships. We also sketch pat-

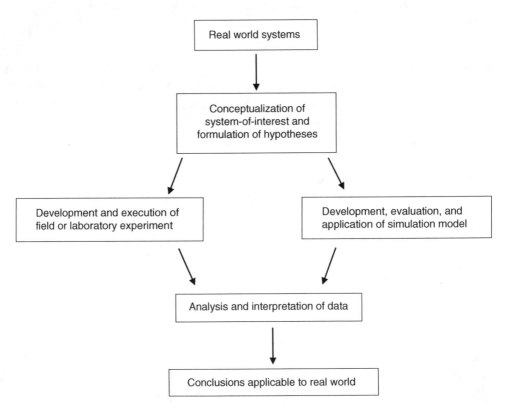

Fig. 2.28 Comparison of simulation to laboratory and field experiments as alternative, complimentary methods of problem solving (modified from Van Dyne, 1969).

COMMON-SENSE SOLUTIONS: THREE EXERCISES

terns of behavior that we expect our model will exhibit, often in terms of the general temporal dynamics of key system components.

In the second phase, we develop a quantitative model of the system-of-interest (Fig. 2.1). This basically involves translating our conceptual model, which is represented diagrammatically and using words, into a series of mathematical equations that collectively form the quantitative model. This translation, or quantification, is based on the consideration of various types of information about the real system. We then solve all of the model equations, or instruct a computer to solve them, each time step over the entire period of simulated time in which we are interested.

2.2.2 The quantitative model (Phase II)

In the third phase, we evaluate the usefulness of the model in meeting our objectives (Fig. 2.1). This process, which is the subject of much debate, commonly is referred to as "model validation" and often erroneously focuses on comparison of model predictions with real system observations as the only validation criterion. We prefer the concept of "model evaluation" based on consideration of a broad array of different aspects of model structure and behavior that make it potentially useful. We may place more emphasis on interpretability of relationships among components within the model or on predictive capabilities of the model, depending on our objectives. Often we are interested in determining how sensitive model predictions are to the uncertainties with which we have represented certain aspects of the model.

2.2.3 Model evaluation (Phase III)

In the final phase, we use the model to answer the questions we identified at the beginning of the modeling project (Fig. 2.1). This involves designing and simulating the same experiments with the model that we ideally would conduct in the real system to answer our questions. We also analyze, interpret, and communicate simulation results using the same general procedures we would use for real-world results. Often, as with real-world experiments, we conduct additional simulated "pilot studies" to explore new ideas that occurred to us as we were analyzing the results of our initial simulated experiments.

2.2.4 Model application (Phase IV)

We now revisit our three problems, noting similarities among our common-sense solutions, and comparing the various steps in our common-sense solutions to the four phases of the systems approach to problem solving described in Section 2.2.

For each of our problems, we first diagrammed the important relationships among components in our system-of-interest, given our questions, and also sketched expected patterns of system behavior. We then quantified, step-by-step, the relationships among these components, checking

2.3 The three problems revisted: the systems approach in theory and practice

the results of our calculations at each step in terms of their reasonableness and usefulness for addressing our questions. Finally, we projected system behavior under appropriate scenarios to address our questions.

This (1) diagram, (2) quantify–calculate–check, (3) project scenarios scheme is generally similar to the (1) conceptual model, (2) quantitative model, (3) model evaluation, (4) model application phases of the systems approach (Fig. 2.1, left-hand and center portions). The main difference is that, rather than proceeding smoothly, one time, through the four theoretical phases, we sometimes cycled through the first three phases more than once (three times in our population example and five times in our Commons example). In practice, we seldom quantify the entire conceptual model before running simulations and evaluating model behavior. We usually construct a simple "running" model as quickly as possible and then expand it gradually through a series of intermediate models, adding just a component or two at a time and checking for errors at each step, until we have quantified the entire model. This procedure helps us identify the inevitable errors we will make during model development.

Thus, in practice, to avoid some of the more common pitfalls associated with model development (Chapter 7), we engage in three general types of modeling activities (Fig. 2.1, right-hand portion). First, we develop a preliminary conceptual model of the entire system-of-interest, as well as a "plan of attack" for quantifying the model piece-by-piece. Second, we construct a series of intermediate developmental models, following our plan of attack, starting with what usually seems a trivially small portion of the overall model. We quantify, "run," and evaluate each model, and make any necessary adjustments, before proceeding to the next one. We add only a tiny piece to each new model, following our plan, which we also may modify as we proceed. We repeat the quantify, run, evaluate, and adjust sequence until we have reassembled the entire model, which seldom is identical to our original conceptual model. And finally, we use the final, "reconstructed," model to address our questions, which we also may have modified as we sharpened our focus during model development.

3
Theory I: the conceptual model

The goal of the first phase of the simulation modeling process is to develop a conceptual, or qualitative, model of the system-of-interest (Fig. 3.1). Based on a clear statement of the objectives of the modeling project, we abstract from the real system those components that must be considered to address our questions. By including these components within our model and excluding all others, we bound the system-of-interest. Next, we categorize model components depending on their specific roles in describing system structure and identify specific relationships among components that generate system dynamics. We then formally represent the resulting conceptual model, usually as a box-and-arrow diagram indicating points of accumulation of material (boxes), such as individuals, energy, biomass, nutrients, or dollars, and routes by which the material flows within the system (arrows). Finally, we describe expected patterns of model behavior, most often as graphs representing changes in values of important variables within the system over time.

In many respects, conceptual-model formulation is the most intellectually challenging phase of the modeling process. The best basis for the many difficult, and often highly subjective, decisions that must be made regarding choice of model components is a thorough familiarity with the real system. Prior modeling experience also is an asset. There are two general approaches to identifying model components. One makes the initial choice of components as simple as possible and subsequently adds critical components that were overlooked; the other includes initially all components that possibly could have any importance and then deletes superfluous ones. Theoretically, the end product of either approach should be a conceptual model that is no more complex than is absolutely necessary to address our interests. As we will see in Chapter 8, in practice, it is better to begin with the simplest model possible.

Abbreviations used to represent the various steps in this phase of the modeling process ($I_a, \ldots, I_f$) correspond to the parenthetical entries scattered through the three problems we presented in Chapter 2 (Section 2.1). You should refer frequently to these problems as you read this chapter with the dual goal of (1) placing a particular aspect of the theory within the context of a commonsensical approach to problem solving, and (2) viewing our commonsensical approach from a theoretical, systems perspective.

3.1 State the model objectives (I_a)

We begin with a clear statement of the purpose of our model in terms of a problem to be solved or a question to be answered. Questions may arise from general observations of a system, as is the usual case in scientific inquiry, or may be imposed by the practical necessity of evaluating proposed management schemes. We also must specify the criteria the model must meet to be considered useful, given its purpose. Criteria may require that the model have theoretically reasonable structure and interpretable functional relationships, that model behavior corresponds well with expected patterns of real-system behavior, and that model projections correspond well with data from the real system. The relative importance of each type of criteria depends on the purpose of the model. Finally, we must describe the context within which we intend the model to operate. Context includes all the restrictive assumptions we must make for the model to be a useful representation of the real system. Obviously, an exhaustive list of these assumptions is impossible, and how to distinguish between those assumptions that can be left implicit (the earth will continue to revolve around the sun) and those that should be stated explicitly (no changes in basic management practices) will remain enigmatic. Nonetheless, we should give serious thought to the manner in which we can most effectively communicate the essence of this context to those who will use the model or its results.

```
Phase I: The Conceptual Model

    Iₐ:  State the model objectives
    I_b:  Bound the system-of-interest
    I_c:  Categorize the components within the system-of-interest
    I_d:  Identify the relationships among the components of interest
    I_e:  Represent the conceptual model
    I_f:  Describe the expected patterns of model behavior
```

Fig. 3.1 Steps within Phase I of the simulation modeling process: the conceptual model.

Because model objectives provide the framework for model development, the standard for model evaluation, and the context within which simulation results will be interpreted, this is arguably the most crucial step in the entire modeling process. Yet, surprisingly, this step usually receives far less attention than its importance warrants.

Often, our initial formulation of an objective is too broad to address directly and thus is of little use in guiding model development. As a general rule, objectives that begin with "to understand . . ." need to be stated more specifically. Consider, for example, the first few paragraphs with which we introduced each of the examples in Chapter 2. In each case (Hunter-gatherer, Section 2.1.1; Population, Section 2.1.2; and Commons, Section 2.1.3), the first paragraph of the problem contains an informal statement of our objectives that provides a general idea of the problem, but does not specify exactly how we will approach the problem; it remains unclear exactly what we will manipulate and exactly what we will monitor. The first paragraph of the solution restates our objectives in specific terms that indicate exactly what we will manipulate and exactly what we will monitor to address our questions.

Bounding the system-of-interest consists of separating those components that should be included within the system-of-interest from those that should be excluded. We do not want an unnecessarily complex model, but likewise we do not want to exclude components that might be critical to the solution of our problem. In the hunter-gatherer example (Fig. 2.2b), the only components we included within the system-of-interest were the food items in the forest, the harvest of food items, and the total harvest. We did not include the production of new food items because we explicitly assumed that there was no production of food (plants and animals) in the forest during the month of November. We did not include the loss of harvested food items, implicitly assuming that the loss, if any, was unimportantly small within the context of our problem. In the population example (Fig. 2.4d), we included only the number of individuals in the population, number of births and deaths, density-dependent birth rates, hurricane-dependent death rates, and hurricanes. We excluded interactions with other species, as well as noncatastrophic weather events (and the list goes on), because we considered them unimportant at the level of detail we needed to adequately address our questions, that is, to adequately estimate probability of extinction. In the Commons example, we included many more components explicitly in our model (Fig. 2.10b), but a complete list of components excluded, as with the hunter-gatherer and population examples, would approach infinity.

Obviously this first step in conceptual-model formulation is highly subjective. Let's reconsider for a moment some other ways in which we

3.2 Bound the system-of-interest (I_b)

might have bounded the system-of-interest related to our population example (Fig. 2.4d). If, for example, we had decided amount of available prey was an important factor affecting per capita birth rate of our population, and that available prey also was affected by hurricanes, we might have conceptualized the system-of-interest as in Fig. 3.2. (We will define formally the symbols we are using in these diagrams in the next section, but for now the figures can be interpreted informally.) If we then had decided amount of available prey, in addition to responding to hurricanes, also is affected by population size (of our target population), we might have conceptualized the system-of-interest as in Fig. 3.3.

Note that Fig. 3.3 does not represent the only way we might have included available prey in our population model. If we wanted to emphasize that available prey did not alter per capita birth rate, but, rather, affected births in some other manner, we might have conceptualized the system as in Fig. 3.4. If we decided changes in available prey could not be

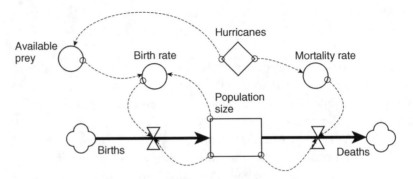

Fig. 3.2 The population model (Fig. 2.4d) modified to represent available prey as a component affecting the per capita birth rate and weather affecting the prey.

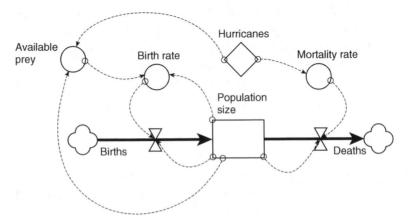

Fig. 3.3 The population model (Fig. 3.2) modified to represent the number of individuals in the population as a system component affecting the availability of prey.

MODELING THEORY

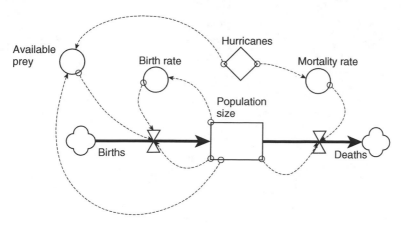

Fig. 3.4 Alternative representation of the population (Fig. 3.3) representing the effect of available prey on births directly, rather than indirectly via birth rate.

represented adequately based solely on the net effect of hurricanes and population size (of our target population), but, rather, we needed to consider density-dependent effects on prey births as distinct from the effects of predation and hurricanes on prey deaths, we might have sketched Fig. 3.5. These alternative representations (Figs 3.3–3.5) differ with regard to what we have chosen to represent explicitly and what we have chosen to leave implicit. None is right or wrong, but one probably is a more useful (clear, concise, informative) view of the system, given our objectives.

Once the system-of-interest has been bounded by separating those components that should be included within the system from those that should be excluded and by identifying particular attributes of system components that are of interest, we proceed to step 3 of conceptual-model formulation: categorizing components within the system-of-interest. System components may not all serve the same purpose in a model. Certainly, they all represent important aspects of the system-of-interest, but there may be as many as seven fundamentally different categories of system components: (1) state variables, (2) material transfers, (3) sources and sinks, (4) information transfers, (5) driving variables, (6) constants, and (7) auxiliary variables (Fig. 3.6) (Forrester, 1961; Innis, 1979; Grant, 1986; Grant et al., 1997).

3.3 Categorize the components within the system-of-interest (I_c)

State variables (Fig. 3.6) represent points of accumulation of material within the system; for example, food items accumulated in the forest, after harvest, accumulated as total harvest (Fig. 2.2b), number of individuals accumulated in a population (Fig. 2.4d), biomass units accumulated as

3.3.1 State variables (□)

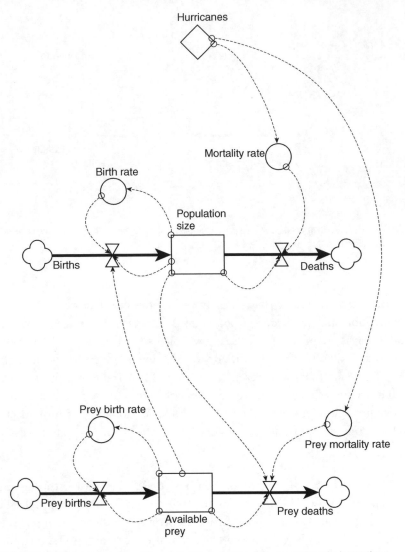

Fig. 3.5 Alternative representation of the population model (Fig. 3.3) representing the interactions between the populations of the focal species (a predator) and the prey species, which now has its own suite of demographic components.

forage on the Commons and biomass units accumulated within an individual animal (Fig. 2.10b). Other examples of state variables might include kcal of energy accumulated in plants, herbivores, and carnivores, if we are interested in energy flow though an ecosystem (Fig. 3.7); and kg of nutrients accumulated in plants, herbivores, carnivores, and an abiotic component, if we are interested in nutrient cycling in an ecosystem (Fig. 3.8).

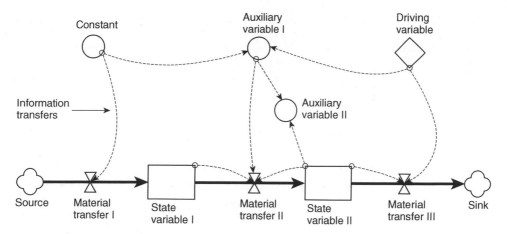

Fig. 3.6 Symbols used to construct conceptual-model diagrams indicating all permissible connections. Thin arrows represent information transfers. Clouds represent sources and sinks.

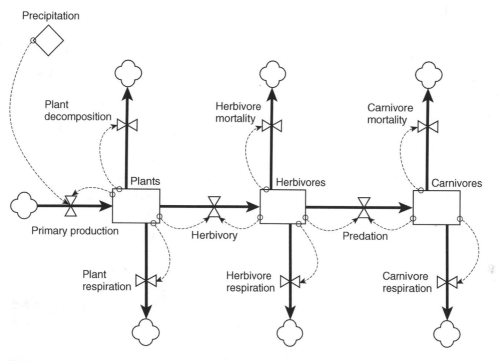

Fig. 3.7 Conceptual-model diagram representing energy flow through an ecosystem.

The accumulation of material in a state variable may increase or decrease via material transfers (Fig. 3.6). A material transfer represents the movement of material (1) from outside the system into a state variable, (2) between two state variables, or (3) from a state variable out of the system; for example, movement of food items from the forest to total harvest

3.3.2 Material transfers (⟶⟶)

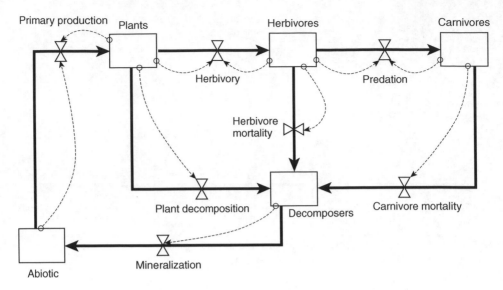

Fig. 3.8 Conceptual-model diagram representing nutrient cycling within a closed ecosystem.

(Fig. 2.2b), movement of individuals from outside the system into a population (births) and from a population out of the system (deaths) (Fig. 2.4d), movement of biomass units from outside the system into forage on the Commons (forage growth) and from forage on the Commons outside the system via grazing loss and nongrazing loss (Fig. 2.10b), and movement of biomass units from outside the system into an individual animal (weight gain) and from an individual animal out of the system (weight loss) (Fig. 2.10b). Other examples of material transfers might include movement of kcal of energy into, through, and out of the plants, herbivores, and carnivores in an ecosystem (Fig. 3.7); and movement of kg of nutrients among plants, herbivores, carnivores, and an abiotic component in an ecosystem (Fig. 3.8).

Note that units of measure associated with material transfers always include the units of measure of the state variable(s) to which they are connected, plus a time unit; for example, number of food items per day ("flowing" from the forest into total harvest), number of individuals dying per year ("flowing" out of the population), units of biomass of forage growth per month ("flowing into standing crop of forage on the Commons), and units of biomass of weight loss per month ("flowing" out of an individual animal on the Commons). Thus two state variables connected by a material transfer must have the same units of measure. Specifying the appropriate time units (second, year, century) occurs in the second phase of systems analysis (II_b), but looking ahead a bit, the basic assumption we make is the rate of material transfer is constant over the time unit we specify. If we feel uncomfortable with this assumption, we need to shorten the time unit. (Discussion of instantaneous and finite rates of change, and differential and difference equations, would be a distraction at this point.)

Sources and sinks (represented by "clouds" in Fig. 3.6) represent origination and termination points, respectively, of material transfers into and out of the system. By definition we are not interested in the level of accumulation of material within sources and sinks, we are interested only in the rates at which material enters and leaves the system; for example, we were not interested in the source of individuals born into our population (Fig. 2.4d), or the source of biomass entering forage or individual animal weight on the Commons (Fig. 2.10b). Likewise, we were not interested in the fate of dead individuals (Figs. 2.4d), or the fate of biomass lost from the Commons via grazing, trampling, and animal weight loss (Fig. 2.10b). We were interested only in the rates at which these material transfers into and out of the systems occurred. Note that closed systems have no sources or sinks; our hunter-gatherer model (Fig. 2.2b) and the model of nutrient cycling within an ecosystem (Fig. 3.8) are examples.

Information transfers (Fig. 3.6) represent the use of information about the state of the system to control the change of state of the system. Control may be direct, such as control of weight loss by maintenance requirements and individual animal weight (Fig. 2.10b); or indirect, such as control of grazing loss by individual animal weight via forage requirements per animal and forage requirement of herd, and control of weight gain by forage on the Commons via relative forage available and weight gain index (Fig. 2.10b). However, although use of the terms "direct control" and "indirect control" sometimes are useful in describing the general structure of a model, note that the idea of "direct control" is a dangerously illusive one if we are referring to control of the dynamics of a complex (highly interconnected) system. Unlike material transfers, the units of measure we use to represent information may change as the information passes from one system component to another, as long as the appropriate conversion factors are used; for example, information about hurricanes (a unit-less index) can be converted into death rate (proportion of population dying per month) (Fig. 2.4d), and information about individual animal weight (units of animal biomass) can be converted, with two intermediate steps, into grazing loss (units of forage biomass) (Fig. 2.10b). We refer collectively to the information transfers within a system as the information network, which connects not only state variables and material transfers but also driving variables, constants, and auxiliary variables (described below).

3.3.5 Driving variables (◇)	Driving variables (Fig. 3.6) affect, but are not affected by, the rest of the system; for example, hurricanes in our population model (Fig. 2.4d) and number of animals in our Commons model (Fig. 2.10b). Not all models have driving variables; our hunter-gatherer model did not. Note that our definition of driving variables is tightly linked to the manner in which we have bounded our system-of-interest. In our energy flow model, we represented precipitation as a driving variable; precipitation affected the system (primary production) but was not affected by any system component (Fig. 3.7). Precipitation is not inherently a driving variable, although we often represent it as such. If we were interested in the dynamics of plant production on a regional scale, perhaps in response to extensive clearing of trees, amount of precipitation entering the system might very well be affected by the amount of forest remaining; shifts in regional rainfall resulting from deforestation are well documented.

3.3.6 Constants (○)	Constants (Fig. 3.6) are numerical values describing important characteristics of a system that do not change, or that can be represented as unchanging, under all of the conditions encountered in any given scenario simulated by the model (any single "run" of the model); for example, maintenance requirements in our Commons model (Fig. 2.10b). As with driving variables, our definition of constants is tightly linked to the manner in which we have conceptualized our system-of-interest. Factors such as environmental temperature and precipitation, which often are represented as driving variables, also may be represented as constants, if they do not change during the course of any given simulation. Note also that we need not, and usually do not, represent all of the mathematical constants in our model as constants in our conceptual model; quite likely there will be coefficients that appear in equations throughout the model that are not, by themselves, conceptually important.

3.3.7 Auxiliary variables (○)	Auxiliary variables (Fig. 3.6) most commonly represent processes or concepts in the system-of-interest that we wish to indicate explicitly, which otherwise would be implicit in the information transfers among model components (constants, driving variables, state variables, material transfers). Such auxiliary variables might be viewed as intermediate steps in determining a rate of material transfer or the value of another auxiliary variable; for example, birth rate and death rate in our population model (Fig. 2.4d). We might have omitted these two auxiliary variables from our conceptual model, connecting hurricanes directly to deaths (Fig. 3.9). Both conceptual models (Figs 2.4d, 3.9) indicate that births are a function of population size and deaths are a function of population size and hurricanes. However, Fig. 2.4d emphasizes that per capita birth rate is a

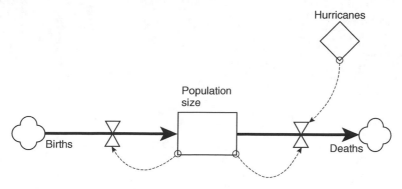

Fig. 3.9 Modified conceptual model of the island population (Fig. 2.4) in which the two auxiliary variables, birth rate and mortality rate, have been eliminated.

function of population size and per capita death rate is a function of hurricanes, presumably providing a more informative conceptual representation of the system, given model objectives. Likewise, we could have eliminated all of the auxiliary variables from our Commons model, thus leaving implicit the details of the connections between forage and animal weight gain, and between individual animal weight and grazing and non-grazing losses (Fig. 3.10a). The goal is to make the conceptual model as useful to the intended audience as possible.

Auxiliary variables also may represent an alternative form of some model component that, although it does not affect system dynamics, is of particular interest to us. Such auxiliary variables often are used simply to express the value of a model component in different units of measure. For example, we might have used an auxiliary variable to represent explicitly total animal biomass on the Commons simply because this was a result of particular interest; note that no information transfers leave total animal biomass (Fig. 3.10b).

3.4 Identify the relationships among the components that are of interest (I_d)

Step 4 of conceptual-model formulation consists of identifying relationships among system components that are of interest. As we have seen in the previous section, there are two ways that system components can be connected: by material transfers or by information transfers (Fig. 3.6). To summarize, material transfers can enter a state variable from a source, connect two state variables, or leave a state variable to a sink. Units of measure of material transfers must be the same as the state variables to which they are connected, with the addition of a time unit; for example, if state variable units are kg/ha, material transfer units could be kg/ha-month or kg/ha-yr. Units of measure of state variables connected by material transfers must be the same. "Information" transferred within the system refers to information about current values of state variables,

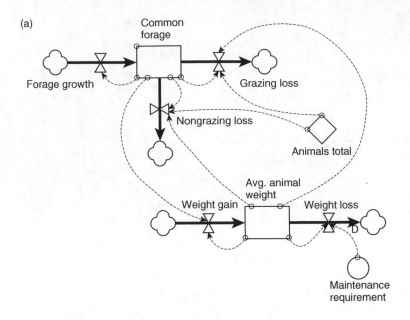

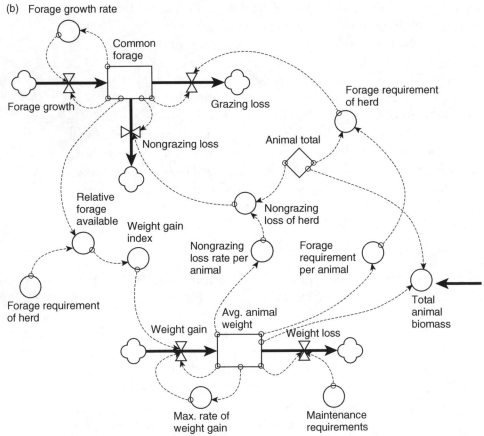

Fig. 3.10 Modified conceptual model of the Commons (Fig. 2.10) in which (a) all auxiliary variables have been eliminated and (b) a new auxiliary variable representing total animal biomass has been added.

driving variables, constants, and auxiliary variables. This information is "transferred" for use in determining the rates at which material transfers occur or for calculating specific results, or "output" auxiliary variables, required of the model. Information transfers can leave state variables, driving variables, constants, and auxiliary variables and can enter material transfers and auxiliary variables. Units of measure of variables affecting a given material transfer or auxiliary variable need not be the same, but, of course, the different units of measure must be manipulated appropriately and must be consistent with the definition of the system components involved.

Often it is useful conceptually to describe larger models in terms of sub-models. The manner in which submodels are defined depends on the particular aspects of the model that we want to emphasize. In general, submodels are defined such that they distinguish important groups of system components and processes. Sometimes submodels are defined to emphasize the different types of material that are "flowing" in the system, with each submodel representing the dynamics of a different material and submodels connected to each other solely via information transfers. We might describe the version of our population model that represents available prey as a state variable (Fig. 3.5) in terms of two submodels, one for our original population, and one for available prey. We might describe the Commons model (Fig. 2.10) as consisting of a forage dynamics sub-model and an individual animal weight dynamics submodel; in fact, we do just that when we formally describe the Commons model in Appendix B. Sometimes submodels are defined such that they distinguish physical environmental, ecological, and social (human) system components, or abiotic and biotic system components. These decisions are completely subjective and are based on whether or not we think division into sub-models facilitates understanding of the model.

3.4.1 Submodels

Formal representation of the conceptual model most commonly takes the form of a box-and-arrow diagram such as those we have been using thus far. As we have seen, such diagrams play an important role in modeling by helping us visualize the "big picture" and by facilitating communication among different people who are interested in a particular system. Although we present this as the fifth step in conceptual-model formulation, and indeed the conceptual-model diagram might be thought of as the end product of the first phase of systems analysis, diagrammatic representation of the conceptual model usually is concurrent with the earlier steps and aids them greatly. Conceptual-model diagrams also provide a framework that facilitates subsequent quantification of the model because equations can be related directly to specific parts of the conceptual model.

3.5 Represent the conceptual model (I_e)

3.5.1 Conceptual-model diagrams

A variety of schemes exists for formal representation of the conceptual model. The origin of the particular symbols we use throughout this book, and the concepts they represent, can be traced back (at least) to the "Forrester Diagrams" used for modeling dynamics of industrial systems (Forrester, 1961). We have modified the shapes of the symbols a bit over the years (Grant, 1986; Grant et al., 1997), but only for the sake of convenience in generating them.

3.6 Describe the expected patterns of model behavior (I_f)

We almost always have some expectations concerning patterns of model behavior before ever running the first simulation. We should formally describe these expectations so we can draw upon them (1) as points of reference during model evaluation and (2) to ensure that the model provides the types of projections that allow us to address our questions directly during model application. These expectations are based on all of our current knowledge about the system-of-interest, including, but not limited to, what can be supported by available data. Most often we know more about relationships among variables within the system-of-interest than can be documented in a rigorous way by data, and almost always there are important aspects of system dynamics for which we simply have no data. We should describe the expected behavior of those variables that most effectively represent this broader knowledge, thus allowing a more extensive evaluation of model behavior than would be possible based solely on data. We also should describe the expected behavior of those variables that most directly represent hypotheses that we want to test. We usually formalize these expectations as graphs representing changes in values of important variables over time; for example, how we expect the number of food items in the forest to decrease (Fig. 2.3), how we expect the population to grow with and without hurricanes (Fig. 2.5a), and how we expect forage biomass to change on the Commons (Fig. 2.11). But we also should note any aspect of system behavior for which we have an expectation. For example, in our hunter-gatherer example, we noted that if we began with 100 food items in the forest, we expected to be able to accumulate the required 75 food items in less than 30 days (Fig. 2.3a). Other examples might include noting maximum and minimum values of system components, and proportional relationships among system components (ratio between A and B, A should go down as B goes up, etc.). The fact that some of our expectations represent the very hypotheses we want to test may seem trivially obvious; theoretically this always should be the case. Unfortunately, all too often, we succumb to the temptation of leaving our expectations implicit and, hence, imprecise, only to find the model incapable of generating some important aspect of system behavior we need to answer our questions.

4

Theory II: the quantitative model

The goal of Phase II of the simulation modeling process is to develop a quantitative model of the system-of-interest (Fig. 4.1). Using the conceptual model as a template for this quantitative development, we describe the rules governing the flow of materials in the model (the dynamics of the system) using mathematical equations. The first step is to choose a general quantitative structure for the model. One useful mathematical format for quantifying the types of conceptual models we have developed in Chapter 3 consists of a set of difference equations, one for each state variable. From one time step to the next the value of each state variable is calculated as the net difference between the material transfers that enter and leave the state variable. The rate of each material transfer is calculated based on information transferred to it from the appropriate constants, driving variables, auxiliary variables, and state variables.

Having decided on the general quantitative structure, we next must develop the specific equations that collectively comprise the model. This consists of choosing the basic time unit for the solution of model equations (e.g., 1 minute, 1 month, 1 year), the functional forms of model equations (e.g., linear, exponential, sigmoid, sinusoidal), and estimating the parameters of model equations. The best type of information we can use to develop model equations is data from the real system. Available data often can be analyzed using standard statistical procedures to quantify various aspects of the model. However, most commonly there will be some aspects of the model for which no data are available and for which we cannot collect new data at the present time. In such cases we may be able to use information based on theoretical or generally applicable empirical relationships. Alternatively we may be able to convert quantitative information, either from the technical literature or from "expert opinion," into a quantitative form. Finally, we may be able to use information generated from experimentation with the model itself to gain insight into the quantification of certain relationships.

The final step in the quantitative specification of the model involve executing the baseline simulation. Baseline simulation refers to solving the model, or simulating behavior of the system-of-interest, under a specific set of conditions that often represent the "normal" situation for the system, or are analogous to the "control" treatment in a designed experiment. Model equations are presented formally by listing them sequentially in some logical order that unambiguously describes how to solve the model.

Abbreviations used to represent the various steps in this phase of the modeling process ($II_a, \ldots, II_e$) correspond to the parenthetical entries scattered through the three problems we presented in Chapter 2 (Section 2.1). You should refer frequently to these problems as you read this chapter with the dual goal of (1) placing a particular aspect of the theory within the context of a commonsensical approach to problem solving, and (2) viewing our commonsensical approach from a theoretical perspective.

4.1 Select the general quantitative structure for the model (II_a)

Theoretically, we should be able to represent the dynamics generated by the relationships among system components equally well in different mathematical formats. That is, results of simulations should not depend on the particular mathematics (e.g., matrix algebra, differential equations) or computer languages (e.g., FORTRAN, C++, Visual Basic) or computer programs (e.g., STELLA® (isee systems, Inc.), VENSIM® (Ventana Systems, Inc.)) we use to represent them; in practice, this is not always the case. A discussion of the variety of different types of mathematical formats suitable for representing models of dynamic ecological systems is beyond the scope of this book (see Gilman and Hails (1997) and Jørgensen and Bendoricchio (2001) for discussions of mathematical formats).

Phase II: The Quantitative Model
  II_a: Select the general quantitative structure for the model
  II_b: Choose the basic time unit for the simulations
  II_c: Identify the functional forms of the model equations
  II_d: Estimate the parameters of the model equations
  II_e: Execute the baseline simulation

Fig. 4.1 Steps within Phase II of the simulation modeling process: the quantitative model.

However, one of the simplest, most flexible mathematical formats consists of a set of difference equations developed within a general compartment-model (box-and-arrow) structure; this is exactly the structure we used in Chapter 3 to represent qualitatively our conceptual models. The basic structural module is a state variable (Section 3.3.1) with a material transfer (Section 3.3.2) entering and/or leaving it. We "solve" the model, that is, we simulate system dynamics, by recalculating the value of each state variable each time step; we calculate the new value of each state variable by adding the value of each material transfer entering the state variable and subtracting the value of each material transfer leaving the state variable. For example, the equation we would solve each time step (Δt) for a state variable (SV) with one material transfer entering (MTE) and two material transfers leaving (MTL1 and MTL2) would be:

$$SV_{t+1} = SV_t + (MTE_t - MTL1_t - MTL2_t) \star \Delta t$$

Note that for difference equations, Δt always is equal to 1; it represents 1 day (like our hunter-gatherer model), 1 month (our Commons model), 1 year (our population model), or 1 other time unit, depending on our choice of the basic time unit for simulation (Section 4.2). Material transfers are calculated based on information transferred (Section 3.3.4) from other parts of the system, including information about state variables, driving variables (Section 3.3.5), constants (Section 3.3.6), and/or auxiliary variables (Section 3.3.7) (Fig. 3.6).

The general strategy is to write a set of equations that determine at selected points in time the value of each driving variable, auxiliary variable, material transfer, and state variable. Collectively, these equations are the quantitative model. To simulate behavior of the system-of-interest, we first specify the initial (at time zero) conditions of the system, including the initial values of all state variables and the values of all constants. For each unit of simulated time, we then "solve" the model by calculating the equations in the following sequence: (1) driving variable equations (if present), (2) auxiliary variable equations (if present), (3) material transfer equations, and (4) state variable equations (Fig. 4.2). We followed exactly this procedure to calculate the dynamics of our hunter-gatherer (Section 2.1.1), population (Section 2.1.2), and Commons (Section 2.1.3) models. We calculated the new number of food items in the forest each day as:

$$Forest\ food\ items_{t+1} = Forest\ food\ items_t - (Harvest) \star \Delta t \quad \text{(eq. 4.1)}$$

and the number of food items in the total harvest each day as:

$$Total\ harvest_{t+1} = Total\ harvest_t + (Harvest) \star \Delta t \quad \text{(eq. 4.2)}$$

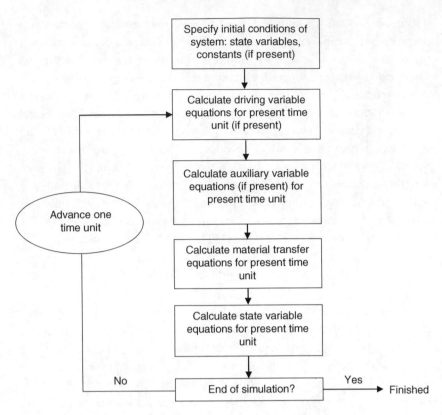

Fig. 4.2 Computing sequence for the model equations within the general compartment-model structure.

the new population size each year as:

$$Population\ size_{t+1} = Population\ size_t + (Births_t - Deaths_t) \star \Delta t \qquad (eq.\ 4.3)$$

and the new common forage and animal weight biomasses each month as:

$$\begin{aligned}Common\ forage_{t+1} &= Common\ forage_t \\ &+ (Forage\ growth_t - Nongrazing\ loss_t - Grazing\ loss_t) \star \Delta t\end{aligned} \quad (eq.\ 4.4)$$

$$\begin{aligned}Average\ animal\ weight_{t+1} \\ &= Average\ animal\ weight_t + (Weight\ gain_t - Weight\ loss_t) \star \Delta t\end{aligned} \quad (eq.\ 4.5)$$

In the following sections, we will take a closer look at some of the material transfer equations in our three examples; all of the equations for hunter-gatherer, population, and Commons models are presented formally in Table 4.1.

Table 4.1 Formal presentation of the model equations for our hunter-gatherer, population, and Commons examples from Chapter 2, organized by the computing sequence of equation types.

Hunter-gatherer (Figure 2.2b):
    Initial conditions:
        Forest Food Items = 80
        Total Harvest = 0
    Material transfer equation:
        $Harvest_t = 0.1 * Forest\ Food\ Items_t$
    State variable equations:
        $Forest\ Food\ Items_{t+1} = Forest\ Food\ Items_t - (Harvest_t) * \Delta t$
        $Total\ Harvest_{t+1} = Total\ Harvest_t + (Harvest_t) * \Delta t$

Population (Figure 2.4d):
    Initial conditions:
        Population size = 100
    Driving variable:
        Hurricanes = 0 or 1 (randomly generated based on estimated frequency of occurrence)
    Auxiliary variables:

$Mortality\ rate_t = 0.5$	if $Hurricanes = 0$
$Mortality\ rate_t = 0.99$	if $Hurricanes = 1$
$Birth\ rate_t = 0.729 - (0.0005714 * Population\ size_t)$	if $50 < Population\ size_t \leq 400$
$Birth\ rate_t = 0.7$	if $Population\ size_t \leq 50$
$Birth\ rate_t = 0.5$	if $Population\ size_t > 400$

    Material transfer equations:
        $Births_t = Birth\ rate_t * Population\ size_t$
        $Deaths_t = Mortality\ rate_t * Population\ size_t$
    State variable equation:
        $Population\ size_{t+1} = Population\ size_t + (Births_t - Deaths_t) * \Delta t$

Commons (Figure 2.10b):
    Initial conditions:
        Common forage = 1000
        Animal weight = 10
        Maintenance requirements = 0.1
        Non-grazing loss rate per animal = 0.02
    Driving variable:
        Animal total (depends on scenario)
    Auxiliary variables:

$Forage\ growth\ rate_t = 1.2 - (0.001 * Common\ forage_t)$	if $200 < Common\ forage_t \leq 1200$
$Forage\ growth\ rate_t = 1.0$	if $Common\ forage_t \leq 200$
$Forage\ growth\ rate_t = 0$	if $Common\ forage_t > 1200$
$Maximum\ rate\ of\ weight\ gain_t = 1.32 - (0.0132 * Average\ animal\ weight_t)$	if $5.0 < Average\ animal\ weight_t \leq 100$
$Maximum\ rate\ of\ weight\ gain_t = 0$	if $Average\ animal\ weight_t \leq 5.0$ (lower lethal limit)
$Maximum\ rate\ of\ weight\ gain_t = 0$	if $Average\ animal\ weight_t > 100$
$Maximum\ rate\ of\ weight\ gain_t = 1.32$	
$Maximum\ rate\ of\ weight\ gain_t = 1.32$	
$Weight\ gain\ index_t = -0.25 + (0.25 * Relative\ forage\ available_t)$	if $1.0 < Relative\ forage\ available_t \leq 5$
$Weight\ gain\ index_t = 0$	if $Relative\ forage\ available_t \leq 1.0$
$Weight\ gain\ index_t = 1$	if $Relative\ forage\ available_t > 5.0$
$Forage\ requirement\ of\ individual_t = 1 * Average\ animal\ weight_t$	
$Forage\ requirement\ of\ herd_t = Animal\ total_t * Forage\ requirement\ of\ individual_t$	
$Non\text{-}grazing\ loss\ rate\ per\ herd_t = Animal\ total_t * Non\text{-}grazing\ loss\ per\ animal$	
$Relative\ forage\ available_t = Common\ Forage_t\ /\ Forage\ requirement\ of\ herd_t$	

    Material transfer equations:

$Non\text{-}Grazing\ loss_t = Non\text{-}grazing\ loss\ rate\ per\ herd_t * Common\ forage_t$	
$Grazing\ loss_t = Forage\ requirement\ of\ herd_t$	if $Forage\ requirement\ of\ herd_t \leq Common\ forage_t$
$Grazing\ loss_t = Common\ forage_t$	if $Forage\ requirement\ of\ herd_t > Common\ forage_t$
$Forage\ growth_t = Forage\ growth\ rate_t * Common\ forage_t$	
$Weight\ gain_t = Average\ animal\ weight_t * Weight\ gain\ index_t * Maximum\ rate\ of\ weight\ gain_t$	
$Weight\ loss_t = Average\ animal\ weight_t * Maintenance\ requirements$	

    State variable equations:
    $Common\ Forage_{t+1} = Common\ Forage_t + (Forage\ growth_t - Non\text{-}grazing\ loss_t - Grazing\ loss_t) * \Delta t$
    $Average\ animal\ weight_{t+1} = Average\ animal\ weight_t + (Weight\ Gain_t - Weight\ Loss_t) * \Delta t$

4.2 Choose the basic time unit for the simulations (II_b)

After the general quantitative structure is chosen, the next step in quantitative specification of the model is to choose the basic time unit for simulations, which is the time interval (Δt) between iterative solutions of model equations. Our choice of the basic time unit depends on the level of temporal resolution needed for (1) addressing our questions and (2) appropriately representing temporal changes in the rates at which processes within the system-of-interest occur. Implicit in choice of the basic time unit is the assumption that all rates within the system remain constant over any given Δt, which follows logically from the fact that we calculate each rate just once per Δt.

Note the basic time unit need not be confined to familiar units such as 1 day, 1 month, or 1 year; Δt may be defined as 12 hours (night, day), 3 months (4 seasons per year), 365.25 days (length of the earth's orbit around the sun), or any other length that allows us to address our questions and represent temporal dynamics of the system adequately. We calculated harvest in our hunter-gatherer model day by day, births and deaths in our population year by year, and all five material transfers in our Commons model month by month. If, for example, we had felt we needed more precision in our projections of harvest of forest food items, we might have chosen a Δt of 12 hours, or even 1 hour. Likewise, we might have chosen a Δt of 6 months had we thought there were important seasonal differences in birth rate or death rate within our population; or we might have chosen a Δt of 1 week had we thought there were interesting management schemes on the Commons that involved harvesting animals a week or two earlier than normal each year.

4.3 Identify the functional forms of the model equations (II_c)

The next step in quantitative specification of the model is choosing the functional forms of the model equations. We now specify whether the general forms of equations representing specific relationships within the model are linear, sigmoidal, exponential, and so on. In our examples, we specified linear relationships between harvest of food items and number of food items remaining in the forest (Section 2.1.1), birth rate and population size (Fig. 2.6), forage growth rate and common forage (Fig. 2.12), maximum rate of weight gain and animal weight (Fig. 2.14), and weight gain index and relative forage available (Fig. 2.18). Thus, mathematically, the general functional form of each of these relationships is the equation for a straight line:

$$Y_t = \beta_0 + \beta_1 \star X_t \qquad \text{(eq. 4.6)}$$

or, more specifically:

$$Harvest_t = \beta_0 + \beta_1 \star Forest\ food\ items_t \qquad \text{(eq. 4.7)}$$

$$Birth\ rate_t = \beta_0 + \beta_1 \star Population\ size_t \qquad \text{(eq. 4.8)}$$

$$Forage\ growth\ rate_t = \beta_0 + \beta_1 \star Common\ forage_t \qquad \text{(eq. 4.9)}$$

$$Maximum\ rate\ of\ weight\ gain_t = \beta_0 + \beta_1 \star Animal\ weight_t \qquad \text{(eq. 4.10)}$$

$$Weight\ gain\ index_t = \beta_0 + \beta_1 \star Relative\ forage\ available_t \qquad \text{(eq. 4.11)}$$

where β_0 (the Y intercept) and β_1 (the slope of the line) are parameters that must be estimated for each equation.

The best information on which to base the choice of functional forms of model equations is data from direct observation or experimentation with the real system, either collected firsthand or from literature. However, almost always there are important relationships for which we have no data. In such cases, we often can draw on theoretical relationships or generally applicable empirical relationships that are appropriate for the situation being modeled. We also may rely on qualitative information from the literature or on opinions of experts in the field to establish assumptions on which to base the choice of functional forms. In some cases, there may be little known in even a qualitative sense on which to base our choice. In such situations, we can gain insight into possible functional forms of a model equation by hypothesizing different functional forms and observing model behavior in response to each. Through such experimentation with the model, we usually can narrow possible choices by excluding those functional forms that produce unreasonable model behavior. Obviously, the number of equations that can be specified in this manner within any single model necessarily is small. The more equations specified by this trial-and-error method, the higher the likelihood that we will obtain apparently reasonable model behavior purely by chance.

4.3.1 Information on which to base the choice of functional forms

When we choose the functional form for a relationship in our model, we may be able to find a single equation that can be parameterized to represent the appropriate curve adequately; in our hunter-gatherer, population, and Commons examples, we chose linear relationships for simplicity. Other useful equations for more complex curves are readily available (e.g., Grant et al. 1997, p. 53), as are logical procedures for selecting appropriate equations (e.g., Spain, 1982, pp. 47 and 341).

Although delving into technical mathematical details would distract us from the main theme of this book, there are a couple of noteworthy philosophical considerations when choosing the equations with which we

4.3.2 Selecting types of equations to represent the chosen functional forms

construct our models. Both deal with the extent to which the results of particular mathematical manipulations help us think about the "real" world. The first concerns the relative merit of an analytical versus a simulation model. Some mathematical models are solvable in "closed form," that is, they have a general analytical solution that applies to all situations the model can represent. Our hunter-gatherer model, as we noted in the footnote in Chapter 2 (Section 2.1.1), is such a model. We might have represented the changes in food items remaining in the forest as:

$$Forest\ food\ items_t = Forest\ food\ items_0 * e^{-0.1*t} \qquad \text{(eq. 4.12)}$$

where e is the base of natural logarithm, t the number of days since harvesting began, and $Forest\ food\ items_0$ is the number of food items in the forest at time zero, that is, initially. Note that there is no need to calculate the day-by-day harvest of food items. If, for example, we want to know the number of food items in the forest after 5.23 days of harvesting, assuming that initially there were 80, we can calculate it directly ($80 * e^{-0.1*5.23} = 47.42$ items remaining). Other mathematical models have no analytical solution, that is, the best mathematicians have not been able to solve them. These models must be solved step-by-step numerically, like the simulation models we are focusing on in this book. Philosophically, the choice between analytical and simulation models often involves deciding whether we sacrifice ecological realism to obtain mathematical elegance (and the analytical "power" that goes with it) or we sacrifice mathematical elegance to include more ecological realism. If the level of detail at which the system-of-interest must be represented to meet our objectives permits use of an analytical model, then we should use an analytical model. However, if the appropriate level of detail requires a model too complex to be represented in an analytically solvable form, then we should use a simulation model. In ecology and natural resource management, we most often need to represent the system-of-interest in a manner too complex for analytical treatment.

A second consideration is the relative merit of a correlative versus an explanatory (or "mechanistic") model. We develop correlative models primarily to describe and summarize a set of relationships, without regard for the appropriate representation of processes or mechanisms that operate in the real system. The goal is prediction, not explanation. A model representing metabolic rate of an animal solely as a function of time of year might be an example. We develop explanatory models primarily to represent the internal dynamics of the system-of-interest with mathematical expressions that, for us, are interpretable in ecological (or other appropriate subject-matter) terms. The goal is explanation through representation of the causal mechanisms underlying system behavior. A model representing the metabolic rate of an animal as a function of body mass, level of activity, reproductive status, and environmental temperature might be an example. Solely explanatory models and solely correlative models obviously form two idealized ends of a continuum, and classification relative to this dichotomy is based as much on the intent of the modeler as on the structure of the model; a model we view as explanatory at one level of detail we might view as correlative at a finer level of detail. For example, we might label each of our models from Chapter 2 (hunter-gatherer, population, and Commons), independent of the number of variables they include, as either explanatory or correlative, depending on how we view each within the context of the problem we are attempting to solve. We will return to this idea of correlative versus explanatory models when we consider criteria for model evaluation in Chapter 5 (Section 5.1).

Information on which to base parameterization of model equations is of the same four general types used to choose functional forms of model equations: (1) quantitative data, (2) information based on theoretical or generally applicable empirical relationships, (3) qualitative information, and (4) information gained from experimenting with the model itself. In fact, the choice of functional forms and the parameterization of model equations for any given model most commonly are based on the same information. Nonetheless, from a theoretical standpoint, we should view the choice of functional forms and the parameterization of model equations as two distinct steps because the former usually has more profound implications concerning ecological interpretations of model structure than does the latter.

4.4 Estimate the parameters of the model equations (II$_d$)

The specific methodologies that we use to estimate parameters of model equations are as diverse as the field of statistics, and, just as we avoided becoming distracted by technical mathematical details above, we want to avoid becoming distracted by technical statistical details. Statistical considerations that arise within the framework of model parameter estimation are identical to those encountered in analysis of data in general, and excellent statistical texts are readily available (e.g., Ott and Longnecker, 2001). For our purposes, the important point is simply that we use those statistical techniques that allow us to interpret appropriately the available data to quantify important relationships within our system-of-interest. Often equations resulting from statistical analyses actually become a part of the simulation model. In our population and Commons examples, we might have used simple linear regression to relate birth rate to population size, forage growth rate to forage on the Commons, maximum rate of weight gain to animal weight, and weight gain index to relative forage

4.4.1 Statistical analyses within the context of simulation model parameterization

available, and the resulting regression equations would have become parts of the respective simulation models:

$$Birth\ rate_t = 0.0729 - 0.0005714 * Population\ size_t \qquad \text{(eq. 4.13)}$$

$$Forage\ growth\ rate_t = 1.2 - 0.0001 * Common\ forage_t \qquad \text{(eq. 4.14)}$$

$$Maximum\ rate\ of\ weight\ gain_t = 1.32 + 0.0132 * Animal\ weight_t \qquad \text{(eq. 4.15)}$$

$$Weight\ gain\ index_t = -0.25 + 0.25 * Relative\ forage\ available_t \qquad \text{(eq. 4.16)}$$

4.4.2 Quantifying qualitative information

In cases where no quantitative data are available, we usually rely on qualitative information from the literature or on opinions of experts to establish assumptions on which to base our estimates of model parameters. In our hunter-gatherer model, we based our representation of the relationship between harvest rate and food items remaining in the forest on the expert opinion of the group of hunter-gatherers (Section 2.1.1). This perhaps seems a much less rigorous procedure than analyzing data, and, indeed, most of us trained in the natural sciences feel much less comfortable quantifying simulation model equations in this manner. However, we would suggest that quantifying qualitative information is not necessarily a less rigorous procedure than data analysis, only less precise. In both cases, our goal is to use those techniques that allow us to interpret appropriately the available information to quantify important relationships within our system-of-interest. We almost surely will make a bigger mistake by excluding important system processes from our model because we have no data than we will by guessing roughly (making a preliminary hypothesis about) how they might be quantified.

Having quantified our initial guesses, we often can make reasonable adjustments to parameter estimates by observing behavior of the model itself. Of course, as with the choice of functional forms, the number of parameters in any single model adjusted by this trial-and-error method must be relatively small or we run the risk of obtaining apparently reasonable model predictions purely by chance. Finally, we should keep in mind that numerical procedures (use of stochastic models, Section 4.4.3; sensitivity analysis; Section 5.4) we use to deal with uncertainty in model parameters work equally well regardless of whether our uncertainty originates from lack of confidence in quantifying qualitative information, or from lack of confidence in accuracy of data.

4.4.3 Deterministic- versus stochastic-model parameterization

One final consideration regarding model parameterization is whether the model is deterministic or stochastic. So far, we have talked in terms of point estimates of variables and transfer rates within models, thus implying that we are dealing with deterministic models. However, the same parameterization procedures just presented are equally applicable to stochastic models except that some additional information is needed to represent inherent randomness of variables occurring within the model.

Any model that contains one or more random variables is, by definition, stochastic. We use a stochastic model if we need to explicitly represent randomness, or uncertainty, in the dynamics of our system of interest to adequately address our questions. This uncertainty may come from randomness we consider inherent within the system or from the uncertainty associated with our estimation of model parameters. Randomness that we consider inherent within the system often comes from driving variables, such as the occurrence of hurricanes in our population model, although such randomness can originate from any system component. Uncertainty also is associated with our estimation of most model parameters; if we determine (often via sensitivity analysis; Section 5.4) the uncertainty associated with certain parameters might change the interpretation of our results, we may choose to represent these model components as random variables, with the degree of their variability reflecting our level of uncertainty in their estimates. As we will see in Section 5.4.1, there also are other ways to deal with this type of uncertainty.

Representation of a random variable in a stochastic model requires specification of either a statistical distribution or an empirical frequency distribution from which values of the variable are selected randomly, rather than specification of a single-point estimate such as is required for deterministic models. In our population model, each year we determined if a hurricane occurred by drawing randomly from a binomial distribution (0 = no hurricane, 1 = hurricane) with the probability (p) of a hurricane occurring depending on the scenario simulated; for example, during the baseline scenario when hurricanes occurred 10% of the time, the yearly probability of a hurricane occurring was 0.1 ($p=0.1$)

4.5 Execute the baseline simulation (II$_e$)

The baseline simulation represents the behavior of the system-of-interest, or the solution of the model, under the particular set of conditions that we wish to use as a benchmark or standard against which to compare changes in system behavior resulting from other sets of conditions of interest to us. The baseline simulation is the end product of the second phase of the simulation modeling process just as the conceptual-model diagram is the end product of the first phase. As we will see in Chapter 5, model evaluation involves close examination of the baseline simulation and comparisons of baseline simulation projections of various system attributes with corresponding attributes observed in the real system. As we will see in Chapter 6, application of the model involves comparison of baseline simulation results with results of simulations representing management policies or environmental situations that we wish to examine. We identified 100 food items in the forest at the beginning of November, population dynamics without hurricanes, and forage and animal weight dynamics with two animals on the Commons as the baseline conditions for our hunter-gatherer, population, and Commons models, respectively.

4.5.1 Baseline simulations for stochastic models

Whereas there is only one baseline simulation for deterministic models, baseline simulation for stochastic models consists of a set of replicate simulations, each representing baseline conditions, but with specific system dynamics differing due to chance variation in one or more random variables in the model. Thus, for stochastic models, an additional consideration arises at this point, namely, how many replicate simulations should be run.

This question is directly analogous to the question of how many samples should be obtained in an experiment in the real world, and the answer depends on (1) the inherent variability of items we are sampling and (2) the magnitude of difference between different samples of items we consider practically significant relative to our project objectives. Thus, if we rephrase the question formally as "How large a sample must we obtain in order to show that a true difference of magnitude Δ between alternative management policies or environmental situations is statistically significant at a significance level α with a probability of β that the difference will be detected if it exists?", then we can draw upon the sample size formula found in basic statistics textbooks to determine quantitatively how many stochastic repetitions to run (e.g., Ott and Longnecker, 2001, p. 315). We can estimate of the variability of items we are sampling by running relatively many stochastic simulations of baseline conditions and calculating the variance of the model output in which we are interested. Although it is difficult to define "relatively many," as we increase the number of these initial simulations, the estimate of variance should reach a stable level. We must determine the magnitude of differences between samples we consider practically important independent of the model, based on our knowledge of what size differences are relevant in the real system given our objectives.

5
Theory III: model evaluation

The goal of Phase III of the simulation modeling process is to evaluate the model in terms of its relative usefulness for a specific purpose. A model that is very useful for one purpose might be useless or, worse, misleading for another purpose. Most commonly, this process is referred to as "model validation," erroneously suggesting that a "valid" or "correct" model exists for any given system-of-interest. Holling (1978), among others, recognized this was an unfortunate choice of terms and suggested we might better refer to the process of "model invalidation," making an analogy between the process of attempting to invalidate a model and the process of attempting to refute a hypothesis via the scientific method. This analogy has merit in that a model can be viewed as a collection of hypotheses that represents our current understanding of the structure and function of the system-of-interest. However, it tends to overemphasize the role of quantitative, often statistical, comparisons of model predictions with selected observations from the real system as the sole criterion for evaluation.

Rykiel (1996) argues convincingly that different validation criteria are appropriate for different types of models and suggests that validation should mean simply a model is acceptable for its intended use. Rykiel also emphasizes that no generally accepted standards currently exist for the validation of ecological models and provides an excellent discussion of the semantic and philosophical debate concerning validation, in which ecological modelers still are involved. Although details of the debate may seem quite confusing, we believe the basic ideas involved in evaluating the usefulness of a model are easy to understand. Thus, we prefer to refer simply to the process of "model evaluation" and to focus our attention on the examination of various characteristics of a model that make it a potentially useful tool.

During model evaluation, we should examine a broad array of qualitative as well as quantitative aspects of model structure and behavior (Fig. 5.1). We begin by assessing reasonableness of model structure and interpretability of functional relationships within the model. Reasonableness and interpretability are defined relative to the ecological, economic, or other subject-matter context of the model. Next, we evaluate correspondence between model behavior and the expected patterns of model behavior that we described during conceptual-model formulation. We then examine more formally the correspondence between model projections and real-system data, if available. These comparisons may or may not involve the use of statistical tests. Finally, we conduct a sensitivity analysis of the model, which usually consists of sequentially varying one model parameter (or set of parameters) at a time and monitoring the subsequent effects on important aspects of model behavior. By identifying those parameters that most affect model behavior, sensitivity analysis provides valuable insight into the functioning of the model and also suggests the level of confidence we should have in model projections. Relative importance of these various steps for any given model depends on the specific objectives of the modeling project.

Abbreviations used to represent the various steps in this phase of the modeling process (III$_a$, . . . , III$_d$) correspond to the parenthetical entries scattered through the three problems we presented in Chapter 2 (Section 2.1). You should refer frequently to these problems as you read this chapter with the dual goal of (1) placing a particular aspect of the theory within the context of a commonsensical approach to problem solving, and (2) viewing our commonsensical approach from a theoretical perspective.

Phase III: Model Evaluation

III$_a$: Assess the reasonableness of the model structure and the interpretability of functional relationships within the model

III$_b$: Evaluate the correspondence between model behavior and the expected patterns of model behavior

III$_c$: Examine the correspondence between model projections and the data from the real system

III$_d$: Determine the sensitivity of model projections to changes in the values of important parameters

Fig. 5.1 Steps within Phase III of the simulation modeling process: model evaluation.

This first step involves attempting to refute aspects of model structure and functional relationships within the model based on their lack of correspondence with the real system. Thus, this step receives particular emphasis with explanatory models and might be omitted for strictly correlative models. Explanatory models are those whose objectives require that relationships within the model be interpretable in ecological (or other subject matter) terms. Correlative models are those whose objectives require only that they project system dynamics adequately, regardless of the specific relationships that generate those dynamics. The procedure is exactly the same as scientific hypothesis testing. The hypotheses tested are our hypotheses about how functional relationships within the model work and about the structural connections among individual parts of the model. Hypothesis tests are based on the best information available about the corresponding aspects of the real system, viewed at the appropriate level of detail. If any aspect of model structure or any functional relationship within the model can be shown to be an inadequate representation of the corresponding aspect of the real system, then that particular portion of the model is refuted. Criteria for failing hypothesis tests may be qualitative or quantitative, depending on objectives of the modeling project and the type of information available from the real system.

This step is inherently subjective and, hence, somewhat difficult to describe precisely. However, we believe the general idea can be demonstrated clearly via our three examples from Chapter 2. In our food harvesting and population examples, it appeared relatively easy to assess the reasonableness of model structure based directly on the problem descriptions. The group of hunter-gatherers harvested food from the forest at a rate that depended on the amount of food left in the forest (Fig. 2.2). Population size changed as the result of births and deaths, with birth rate depending on current population size and death rate depending on occurrence of hurricanes (Fig. 2.4d). These are not the only conceptual models that we might have used to represent these systems, but it would be difficult to argue that they are unreasonable.

In our Commons example, it was more difficult to assess the reasonableness of model structure because the relative complexity of the system-of-interest provided the opportunity to develop a variety of useful models of the system. The text and figures in Section 2.1.3 associated with development of the conceptual model (Figs 2.8, 2.9, 2.10, and the text describing those figures) offer some insight into the thought processes we used to assess the reasonableness of model structure. (As we will see in Chapter 8, in practice, development of the conceptual model and assessment of the reasonableness of model structure are simultaneous activities, and the process requires a pencil in one hand and an eraser in the other.)

Evaluating the interpretability of the functional relationships in all three of these models, arguably, appeared relatively straightforward. It

5.1 Assess the reasonableness of the model structure and the interpretability of functional relationships within the model (III$_a$)

makes sense that rate of food harvest would decline as food became less abundant and, hence, more difficult to find (Fig. 2.2b). It also makes sense that birth rate would decline as population size increases (Fig. 2.6), due to increased competition among animals for limited resources. Similarly, it makes sense that forage growth rate would decline as forage biomass on the Commons increased (Fig. 2.12), due to increased competition among plants for limited resources, and that rate of animal growth would decline as body weight increased (Fig. 2.14) toward a maximum level. However, implicit in these evaluations of interpretability is identification of the level of interpretability we deem necessary to meet project objectives. That is, based on project objectives, where along the scale from purely correlative to purely explanatory should the model we have developed be positioned.

Of course, a model can not be positioned along the correlative–explanatory scale in any absolute sense. We know from general systems theory that there always are less detailed, less explanatory representations of the system-of-interest, and more detailed, more explanatory representations. Thus, the same model might be viewed as correlative from one perspective and explanatory from another. Nonetheless, through the modeling process we will have identified at least an approximate "benchmark" on the correlative–explanatory scale against which we should evaluate the interpretability of functional relationships in our model. For example, in the Commons model, almost surely we would have thought the representation of forage growth solely as a function of month of the year was inappropriate because it lacked sufficient detail. Representation of forage growth as a function of current forage biomass seemed essential since forage biomass is affected by grazing pressure, and response to different levels of grazing pressure was a principal focus of the project. On the other hand, almost surely we would have thought the representation of forage growth as a function of soil water content, level of solar radiation, and surface area of leaves was inappropriate because it contained too many details. Background information on the system seemed to describe forage dynamics quite well without reference to these details.

5.2 Evaluate the correspondence between model behavior and the expected patterns of model behavior (III_b)

During the second step in model evaluation, we compare model behavior to our *a priori* expectations, which we described during the final step in conceptual-model formulation. In our hunter-gatherer example, we compared our calculations of how the number of food items in the forest would decrease during November, assuming we began with 100 items, to our expectations based on how the number of food items typically had been observed to decrease during November (Fig. 2.3). In our population example, we compared our calculations of population dynamics with the historic occurrence of hurricanes to our expectations based on descriptions of historical population fluctuations (Fig. 2.5). In our Commons example, we compared our calculations of long-term (10-year) forage dynamics with two grazing animals on the Commons to our expectations based on descriptions of historical forage dynamics (Fig. 2.20a). We also compared our expectations of final animal weights to those historically observed (Fig. 2.20b). In these comparisons, we are looking for obvious impossibilities or implausibilities in the baseline simulation, such as negative values for state variables or material transfers that by definition must be positive, or impossibly high or low values for state variables or material transfers. In addition to examining the baseline simulation for unreasonable behavior, we might also examine model behavior over a wide range of input conditions (wide range of values for driving variables) to attempt to expose additional inadequacies in the model. It may seem strange to mention such obvious shortcomings in model behavior, but models that initially show gross inconsistencies with our expectations based on general knowledge about the real system are very common. At this point, these gross inconsistencies may result from a fundamental misconception about the nature of relationships within the system-of-interest, in which case changes in the conceptual model are required. Or, they may result from erroneous *a priori* expectations, which we only now identify as erroneous based on a better understanding of system processes gained from the model itself, as was the case with our initial expectations of animal growth in our Commons model (Fig. 2.15).

In our Commons model, we initially expected that animal weights would increase linearly, however, our calculations indicated a sigmoid growth form (Fig. 2.15). We were thus obliged to reconcile this difference. Either our calculations were flawed, our expectation was flawed, or both. In this case, after reexamining the relationship between the maximum rate of weight gain and animal weight, and recalling our animals always grow proportionally faster when they are small, we decided our expectation of linear growth was flawed.

When a model no longer exhibits obviously implausible behavior, we turn our attention to a closer examination of model components: state, driving, and auxiliary variables; constants; and material transfer rates. General dynamics of each component should be reasonable in terms of the timing of maximum and minimum values, the relative amplitude and periodicity of fluctuations, and its relationship to dynamics of other components. Both state variables and material transfers should vary in the right direction and by the right order of magnitude in response to changes in the values of driving variables. Inadequacies detected as a result of this closer examination of model components still may be caused by a fundamental misconception about the nature of relationships within the system-of-interest, but at this stage, it is likely that inadequacies are caused by inclusion of erroneous parameter estimates or perhaps by inclusion of incorrect functional forms of equations.

The third step in model evaluation focuses more specifically on the correspondence between model projections and real-system data. This step receives particular emphasis with correlative models and theoretically might be omitted for strictly explanatory models, although in practice we almost always are interested to some extent in a model's ability to project system dynamics. Strictly speaking, data from the real system that are used in model evaluation must be independent of data used to develop the model. If the same data used in the quantitative-model specification are used in model evaluation, we can hardly expect to reject the model, because we would have already examined and failed to reject any aspects of the model that were quantified based on those data. The situation is analogous to quantifying a regression model using a given data set, concluding that we will tentatively accept the regression model because it has an r of 0.90 and then "evaluating" the model simply by observing that it does indeed project values in that given data set well. The appropriate evaluation for the regression model is to use the model, as parameterized based on the first data set, to predict values of the dependent variable in a new data set based on values of the independent variables in that new data set. Likewise, for this step in simulation model evaluation to be meaningful, model projections must be compared to real-system data that were not used directly to quantify model equations.

Related to the need to use independent data to evaluate a simulation model, it is important to realize that we use different types of data in different ways during simulation model development. Commonly, data naturally sort themselves into three general categories: those used to (1) evaluate model behavior, (2) quantify driving variables, and (3) quantify auxiliary and material transfer equations (and associated constants). Data we use to evaluate the model often are time series of values of state variables; for example, we used the time series of values of forage biomass in each treatment during each month of the grazing experiment (Figs 2.13, 2.19) to evaluate the final Commons model. Data we use to quantify driving variables often are time series of values over the same historic period for which we have the time series of state variable values; for example, we used the time series of values of the number of animals in each treatment during each month of the grazing experiment to "drive" our Commons model. When we evaluated the Commons model (Fig. 2.10b), the number of animals in each treatment actually was constant from month to month, but it is easy to imagine a treatment in which we changed the number of grazing animals each month; for example, the time series representing the number of grazing animals during each of the 12 months might have been 1, 1, 2, 2, . . . 6, 6, instead of 1, 1, . . . 1, 1. Data we use to quantify auxiliary and material transfer equations are of various types, including time series data, but the distinguishing feature is these data are analyzed (manipulated) in some appropriate way and the results of these analyses (manipulations) are used to write model equations. For example, if we had had actual data relating forage growth rate to current forage biomass on our Commons, we might have used simple linear regression to obtain the equation for the line in Fig. 2.12, which we subsequently used to quantify the auxiliary variable for per capita forage growth.

Thus, ideally, we would have a group of data we could analyze to quantify each of the auxiliary and material transfer equations in the model, and time series data over the same time period for each state variable and each driving variable. We then could compare simulated to historic system behavior by "driving" or "forcing" the model with the time series of driving variable values and comparing the resulting time series of simu-lated values for each state variable, which were generated by the model equations, to the time series of data for each state variable. We also could compare any single simulated value to its historical counterpart – for example, had we had the appropriate "real-system" data, we might have compared simulated and observed, minimum population size, maximum forage biomass, or final animal weight.

The manner in which we examine correspondence between model projections and data from the real system depends on the specific objec-tives of our project, the type of model we have developed, and the type of data available. If we are interested in comparing values projected with a deterministic model to nonreplicated data from the real system, we simply compare the simulated and observed values, nonstatistically, and assess their relative difference or similarity within the context of our problem. If we have a stochastic model and/or replication in data from the real system, our comparisons might include statistical tests of signifi-cance. We already should have determined the number of replicate sto-chastic simulations we need to run at the time we executed the baseline simulations (Section 4.5.1).

Consideration of specific statistical tests would distract us from our central theme. However, one point worthy of mention here is that time series of values projected by a simulation model are auto-correlated. This results from the fact that the state of the system at $t+1$ depends explicitly on the state of the system at time t. This auto-correlation among the time series of values generated for any given system component during a single simulation violates the assumption of independence of observations required by many parametric statistical tests. In any event, the same restrictions that apply to use of statistical tests on real-system data also apply to model projections (simulated data) (see Ott and Longnecker, 2001, for examples of appropriate statistical tests).

5.3.1 Quantitative versus qualitative model evaluation	We should consider for a moment the relative merits of qualitative as opposed to quantitative methods of model evaluation. The tendency is to think of quantitative methods, particularly those involving statistical tests, as being more rigorous and definitive than qualitative methods. We should use quantitative methods of evaluation whenever appropriate. However, we also should keep in mind that objective quantitative methods ultimately rest on subjective judgments relative to the appropriateness of their use in the evaluation procedure. Even statistical hypothesis tests rely on subjective judgment to determine appropriate significance levels for a given problem. Thus quantitative methods are not inherently better than qualitative methods, and both should be viewed within a framework that emphasizes evaluation of the ability of the model to meet project objectives.

5.4 Determine the sensitivity of model projections to changes in the values of important parameters (III$_d$)

Step 4 in model evaluation is to perform a sensitivity analysis on the model. The objective is to determine the degree of response, or sensitivity, of model behavior to changes in various model components (Smith, 1973; Steinhorst, 1979). The basic procedure is to alter the value of one parameter at a time by a specified amount throughout an entire simulation and observe the subsequent effect on selected aspects of model behavior. This procedure may be modified to include varying sets of parameters at the same time and also may include varying the functional forms of relationships within the model. By identifying those parameters or relationships to which model behavior is most responsive, or sensitive, sensitivity analysis provides an indication of the relative accuracy with which each parameter or relationship ideally should be estimated. This information is useful in determining the degree of confidence that should be placed in the model based on the confidence with which the most influential parameters or relationships have been estimated, and also is useful in establishing future research priorities.

We design the sensitivity analysis by specifying the parameters or relationships altered, the manner in which they are altered, and the specific aspects of model behavior monitored. Details of the design depend on objectives of the modeling project, but commonly include altering selected constants and auxiliary variables (usually those that represent functional relationships) over a range of values representing our degree of uncertainty in their true values, and monitoring selected state variables, material transfers, and auxiliary variables (usually those that convert model "output" into a particularly useful form). A possibility with relatively small models is to alter each parameter and each combination of parameters over many values within the selected ranges. However, simulating all possible combinations of even relatively few levels of relatively few parameters quickly becomes an overwhelming task. Most commonly we restrict sensitivity analysis to relatively few parameters of particular inter-est. We decide if we will use statistical tests to compare model sensitivity to changes in one parameter with model sensitivity to changes in another parameter based on the same general considerations described for com-parison of model projections to real-system data (Section 5.3).

In our Commons model, we conducted a sensitivity analysis to examine the effect of our uncertainty in estimating trampling rate (nongrazing loss of forage) on our ability to simulate forage and animal weight dynamics. This uncertainty was of particular concern because the scenarios we wanted to simulate represented the addition of quite a few more animals than had historically been placed on the Commons. Thus, we decided to simulate long-term (10-year) forage and animal weight dynamics assum-ing trampling rates of 5% and 0% per month, in addition to 2% percent per month (our initial estimate), both with 2 animals on the Commons and with 10 animals on the Commons. Results of sensitivity analysis indicated that forage biomass and animal weight dynamics were relatively insensitive to changes in trampling rates with only 2 animals on the Commons (Fig. 2.21), but were very sensitive to changes in trampling rates with 10 animals on the Commons (Fig. 2.22). These results not only suggested that we should give high priority to obtaining more reliable estimates of trampling rate, but also quantitatively related our range of uncertainty in trampling rate estimates to a range of uncertainty in our projections of forage and animal weight dynamics.

Interpretation of sensitivity analysis results differs somewhat from inter-pretation of results obtained in earlier steps of model evaluation. In earlier steps (III$_a$–III$_c$), we attempt to refute the model as being useless. Through sensitivity analysis, we attempt to evaluate more clearly our level of con-fidence in the ability of the model to address our questions. If we have failed to refute the model during the first three steps of the evaluation procedure, we must consider the model irrefutable based on current knowledge about the real system. However, this does not necessarily mean that we have great confidence in the model's ability to answer our questions. Model behavior may be sensitive to changes in some parame-ters or functional forms of equations that we have estimated based on inadequate information. Our lack of confidence in estimates of these influential parameters or functional forms translates into a lack of confi-dence in model projections.

One way to deal with this lack of confidence in model projections due to uncertainty in parameter estimates is to carry out our model applica-tion simulations (Chapter 6) in parallel with different versions of the model, each with a different estimate for the given parameter, or pa-rameters. Another option for dealing with this type of uncertainty is to

5.4.1 Interpreting sensitivity analysis within a model evaluation framework

represent the "uncertain" parameters as random variables (Section 4.4.3), with the degree of their variability reflecting our level of uncertainty in their estimates. Neither option is inherently better than the other; our choice depends, as always, on our specific objectives. In our Commons example, we decided to carry out our model application simulations in parallel with different versions of the model that appropriately reflected the degree of our uncertainty in our estimate of trampling rate. We used three versions of our model, with trampling rate estimates of 0%, 2%, and 5% per animal per month, respectively, to project the various future scenarios in which we were interested.

6

Theory IV: model application

The goal of the final phase of the simulation modeling process is to meet the objectives that were identified at the beginning of the modeling project (Fig. 6.1). Most often we wish to use the model to simulate system dynamics under alternative management strategies or environmental situations. The general scheme for model application exactly follows the steps involved in addressing a question through experimentation in the real world. We first develop and execute the experimental design for simulations, just as we would develop and execute a field or laboratory experiment. Next, we analyze and interpret simulation results, again, using the same analytical tools and interpretive schemes we would use with results from the field or laboratory. Finally, we communicate our results to the appropriate audience, often using the same media we use to report results of field and laboratory studies.

Abbreviations used to represent the various steps in this phase of the modeling process (IV$_a$, . . . , IV$_c$) correspond to the parenthetical entries scattered through the three problems we presented in Chapter 2 (Section 2.1). You should refer frequently to these problems as you read this chapter with the dual goal of (1) placing a particular aspect of the theory within the context of a commonsensical approach to problem solving, and (2) viewing our commonsensical approach from a theoretical perspective.

6.1 Develop and execute the experimental design for the simulations (IV$_a$)

The same principles that apply to the design of experiments conducted in the real world (Cochran and Cox, 1957) apply to experiments conducted, or simulated, on the computer. Our *a priori* expectations, which we described during the final phase of conceptual model formulation, should provide a good guide for experimental design, since some of these

Fig. 6.1 Steps within Phase IV of the simulation modeling process: model application.

expectations represent the hypotheses we want to test. We must avoid the temptation to abandon a well-planned experimental design in favor of the apparent expediency of a "shotgun" approach made possible by the tremendous computing capabilities of modern computers. The ability to generate voluminous results does not preclude the need for a systematic, logical approach to the problem. If our model is stochastic, as part of the experimental design, we also must specify the number of replicate simulations of each experimental treatment we will run. Often this is the same "sample size" we determined when we executed the baseline simulations (Section 4.5.1) and used when we compared model projections to real-system data (Section 5.3). In any event, we determine the required number of replicate simulations just as we would determine sample size for field and laboratory experiments (Ott and Longnecker, 2001).

The experimental design for our hunter-gatherer model consisted of a single scenario in which we simulated food harvest over a 30-day period (Section 2.1.1). The original experimental design for our population model consisted of three treatments (three sets of 100 replicate stochastic simulations) representing scenarios with annual hurricane probabilities of 0%, 10%, and 50%, respectively, which we subsequently extended to include three additional treatments (three additional sets of 100 replicate stochastic simulations) representing scenarios with annual hurricane probabilities of 20%, 30%, and 40%, respectively (Section 2.1.2). Each replicate of each treatment (each simulation) lasted 50 years. The original experimental design for our Commons model consisted of two treatments (two simulations) representing scenarios in which we (1) increased the number of animals we put on the Commons at the same rate our neighbor did and (2) continued to put just one animal on the Commons each year, which we subsequently extended to include one additional treatment representing a scenario in which we added the same number of animals as our neighbor, but we both harvested our animals in June rather than December (Section 2.1.3). Each treatment (each simulation) lasted 120 months. Based on results of our sensitivity analysis of the Commons model (Section 2.1.3), we decided to execute this same experimental design three times, using three versions of our model, each with a different estimate of an important parameter (trampling rate) which we were unable to estimate with confidence.

As was the case with the development of the experimental design for simulations, the same principles that apply to analysis of real-world experimental data apply to analysis of simulated data. Before we compare our model projections under the different simulated treatments to each other, we first compare them to our *a priori* expectations, which we described during the final phase of conceptual model formulation. This comparison represents one last evaluation of the model. Just as we may lose confidence in real-world experimental data if we can not explain aberrant values, we may lose confidence in simulated data if we can not reconcile differences between these data and our expectations. However, since we are projecting system dynamics under conditions that, presumably, have never been observed in the real world, we should be reluctant to discard model projections as useless at this point. The manner in which we interpret differences between how we imagine (hypothesize) the real world functions and actual observations of the real world remains problematic, and continues to provide excellent fodder for philosophical debate. Our point is simply that we analyze and interpret simulated data exactly as we would experimental data from the field or laboratory (Fig. 2.28).

Assuming we decide to proceed with our analysis, we decide how we will compare results from our various simulations (experimental treatments) the same way we would decide how to compare data from the real system obtained from different experimental treatments. In both cases, our decision is based on our project objectives, the details of our experimental design, and on the type of data we have generated. Again, as we discussed within the context of model evaluation (Section 5.3), the same restrictions that apply to use of statistical tests on real-system data (e.g., see Ott and Longnecker, 2001) also apply to model projections (simulated data).

We analyzed the simulated data from our hunter-gatherer and Commons models nonstatistically. For our hunter-gatherer model, we interpreted our simulation results in terms of the difference between the number of days we projected it would take the hunter-gatherers to harvest the food they required (27 days) and the number of days they likely would have to harvest the food (30 days) (Section 2.1.1). For our Commons model, we interpreted our simulation results in terms of the temporal trends in forage biomass and final animal weights under the different scenarios (experimental treatments) (Figs 2.23–2.27). For our population model, we analyzed our simulated data statistically, calculating the probability of population extinction under each hurricane scenario (experimental treatment) (Tables 2.1b,c).

The final step in model application involves communication of simulation results. Within a research setting, this usually means publication in an

6.2 Analyze and interpret the simulation results (IV$_b$)

6.3 Communicate the simulation results (IV$_c$)

appropriate scientific journal. Within a management framework, this implies effective communication of model results to those managers and policy makers whose decisions ultimately impact natural resources. As with the design, analysis, and interpretation of simulated experiments, the communication of simulation results is no different than communication of results of real-world experiments. We must describe the problem we are addressing, our specific objectives, the information base we are drawing upon, the technical method used to analyze the information (that is, we must formally describe the model), the results of evaluation procedures and (simulated) experiments, and our conclusions. We present example scientific reports describing our hunter-gatherer, population, and Commons models in Appendix B.

Communication with potential users of the model is facilitated greatly by their early involvement in the modeling project. Model development usually requires many subjective decisions that, when viewed en masse by users for the first time only after completion of the model, can be quite difficult to explain. User confidence in the model, as well as the overall quality of the model, almost always is higher as the result of early and continued dialogue with potential users. Communication also is easier if emphasis is placed on interpreting general trends in model behavior in ecological terms. General trends usually are of more relevance in a management context than are specific numerical values of model projections. Users and modelers also have more confidence in the model's ability to predict trends and feel less comfortable with any given single numerical projection. However, for most modelers, there is a tendency even with relatively simple models to become preoccupied with presenting detailed results at the expense of a clear overview. This tendency is ironic considering that an underlying rationale for use of a modeling approach is to attain a holistic view of our problem.

7
Some common pitfalls

In this chapter, we point out some of the pitfalls commonly encountered during model development. Although many, if not most, of these pitfalls may sound almost trivially silly when put into words, we find ourselves continually pointing them out to beginners. And, truth be known, we still find ourselves at the bottom of one of these pits from time to time, wondering why we fell. In fact, the majority of the pitfalls we note below might well qualify as "historic landmarks" on the modeling landscape, since most were identified over four decades ago by Jay W. Forrester in his classic work on modeling industrial dynamics (Forrester, 1961). Thus, following humbly in the footsteps of Forrester, as a compliment to our practical "how to" guide, which we present in the next chapter, we offer the following "how not to" guide.

We have organized these pitfalls according to the theoretical phase (or phases) in which we commonly encounter them (Table 7.1).

7.1 Phase I pitfalls: the conceptual model

Step I_a – State model objectives

P1. Inadequate definition of model purpose We should design our model to address specific questions. If our questions are stated vaguely, our decisions regarding what to include in the model will be inconsistent and model structure will be flawed. Model objectives beginning with "To understand . . ." almost always lead to trouble.

P2. Implicit criteria for model evaluation We should specify precisely the criteria our model must meet to be considered useful. If our evaluation criteria remain implicit, and hence vague, our statements regarding the usefulness of the model will be equally vague and will lack credibility.

Common pitfalls encountered during the four theoretical phases of model development.

	Theory												
	I: Conceptual model						II: Quantitative model					III: Mod evaluat	
	a	b	c	d	e	f	a	b	c	d	e	a	b
uate definition of model purpose	√												
t criteria for model evaluation	√												
cription of model context	√												
l choice of scale for the system-of-interest		√											
on of too many components		√											
ss categorization of system components			√										
on of excessive detail				√									
on of circular logic				√									
f precision in conceptual model diagram					√								
ance to make initial hypotheses about system behavior						√							
on of inappropriate mathematics							√						
on of inappropriate software							√						
of inappropriate time unit for simulations								√					
uction of mathematical descriptions without verbal descriptions									√				
stimation of the importance of graphical representations									√				
functional relationships that are not interpretable									√				
ss definition of dimensional units of model components										√			

ents without meaning to obtain dimensional consistency	√				
use qualitative information	√				
move a functional relationship due to lack of data	√				
utomated model parameterization	√				
sophisticated equations	√				
on of the importance of negative feedback and time lags	√				
ition of baseline conditions		√			
utomated solutions to mathematical and programming problems		√			
on of the importance of qualitative aspects of evaluation			√		
conceptually flawed functional relationships			√		
rejection of surprising model behavior without explanation				√	
interpretation of the initial phase of model behavior				√	
utomated model evaluation					√
elieve all data from the real system are correct					√
interpretation of statistical tests used in model evaluation					√
n of sensitivity analysis					
quate model sensitivity with model inadequacy					
imental design for model application					
erestimate the range of model applicability					
interpretation of statistical tests used in model application					
municate numerical results directly in ecological terms					
ply false expectations regarding model projections					

P3. No description of model context We should describe the context within which we intend the model operate. If we do not list the restrictive assumptions regarding the real-world conditions under which our model is useful, our model may be applied inappropriately and provide erroneous projections. A frequent complaint of modelers is that their model appears flawed because it has been applied to problems it was not designed to address. Failure to describe model context invites such misuse.

Step I_b – Bound the system of interest

P4. Casual choice of scale for the system-of-interest We should give careful thought to the manner in which we bound the system-of-interest in time and space. If we bound the system too narrowly, we risk losing the possibility to simulate those situations that would be most helpful in solving our problem. If we bound the system too broadly, we risk focusing too much attention on irrelevant components and processes, and failing to recognize the interactions most important for solving our problem. For example, we probably could not adequately represent the effects of global climate change on land cover in the British Isles by focusing on 1 hectare of land over a period of 1 year, or by focusing on shifts in the earth's orbit around the sun over a period of 1 billion years.

P5. Inclusion of too many components We should resist the tendency to include in the model every component that comes to mind as we think about our problem. The criteria for including a component should be equally rigorous as the criteria for exclusion. Too many components can hinder our recognition and appropriate representation of important interactions.

Step I_c – Categorize components within the system-of-interest

P6. Careless categorization of system components We should be rigorous in our categorization of components within the system-of-interest. Each type of component plays a different role in model structure. Components that appear similar in the real system need not necessarily play similar roles in terms of model structure. For example, representing components as state variables when they may be better represented as auxiliary variables; a model addressing certain questions about predator population dynamics might appropriately represent both predator and prey populations as state variables, whereas a model addressing different questions about predator population dynamics might appropriately represent the predator population as a state variable and the prey population as a driving variable (assuming changes in the prey population were not affected). This along with a lack of precision in the conceptual model diagrams (P9) are two of the most common pitfalls for beginning modelers.

Step I_d – Identify the relationships among the components of interest

P7. Inclusion of excessive detail We should resist the tendency to represent components and processes in our model with great detail, just because we have the knowledge to do so. Too many details can obscure important cause–effect relationships that operate at more aggregated levels of organization; we lose sight of the forest because we can not see past the trees.

P8. Inclusion of circular logic We should scrutinize the conceptual model continually in search of circular logic. As connectedness of the model increases, we risk adding circular connections (via information transfers) among auxiliary variables and material transfers; that is, we represent A as a function of B, B as a function of C, and C as a function of A.

P9. Lack of precision in conceptual model diagram We should familiarize ourselves with the precise definitions of the types of variables (state variable, driving variable, material transfer, etc.) and use symbols with precise definitions to construct our conceptual models. Imprecise description of the conceptual model greatly increases the likelihood of categorizing system components carelessly (P6) and of including circular logic in model structure (P8). It also greatly increases the likelihood of subsequent errors during model quantification. This along with carelessly categorizing system components (P6) are two of the most common pitfalls. Beginners should spend ample time becoming familiar with the definitions for each variable type before they begin conceptualizing their system of interest.

Step I_e – Represent the conceptual model

P10. Reluctance to make initial hypotheses about system behavior We should formulate initial written hypotheses about all aspects of the behavior of the system-of-interest. We should avoid the tendency to view these *a priori* hypotheses as the answers the model will give us; rather, we should view them just as we view the hypotheses we make before any experiment. Our initial written hypotheses, which often are accompanied by graphical representations, serve as important heuristic reference points during subsequent model development and evaluation.

Step I_f – Describe expected patterns of model behavior

P11. Selection of inappropriate mathematics We should choose the mathematics we use to represent the model carefully (or consult with an experienced simulation modeler). Not all mathematical formats lend themselves equally well to the representation of a given problem. We risk the solution of the model becoming a mathematical exercise in which some steps are devoid of ecological interpretation, thus reducing the credibility of model results.

7.2 Phase II pitfalls: the quantitative model

Step II_a – Select the general quantitative structure for the model

P12. Selection of inappropriate software We should choose the software we use to program the model carefully (or consult with an experienced simulation modeler). This is closely related to choice of the mathematics we use to represent the model (P11). Not all computer languages and simulation programs lend themselves equally well to the representation of a given problem. We risk the solution of the model becoming more influenced by constraints of the computer language or simulation program than by the ecological relationships we are trying to represent, thus reducing the credibility of model results.

Step II$_b$ – Choose basic time unit for simulation

P13. Choice of inappropriate time unit for simulations We should use great care in selecting the basic time unit for our simulations. If our time unit is too long, we risk violating the assumption that all rates of change in the system can be viewed as constant during any given time unit. This can preclude the possibility of appropriately representing negative feedback within the system (P23), which can produce artificial instability in the model. If our time unit is too short, we risk reducing the interpretability, and, hence, credibility, of the model, and we also will increase the length of time needed to run simulations unnecessarily.

Step II$_c$ – Identify the functional forms of the model equations

P14. Construction of mathematical descriptions without clear verbal descriptions We should describe functional relationships clearly in words before we attempt to describe them mathematically. If we underestimate the importance of verbal descriptions, we risk being precisely wrong in our mathematical formulations. That is, we should use the precise, but inflexible, language of mathematics to "fine tune" our general, flexible verbal descriptions. Our initial written hypotheses about system behavior (P10) should provide a good point of departure for our verbal descriptions. We have found that a good rule of thumb is to try to explain the model verbally within a couple of minutes. If there are areas of the model that are difficult to explain, they will more than likely be difficult to model mathematically. These are generally the areas that require more research and thought.

P15. Underestimation of the importance of graphical representations We should describe functional relationships graphically whenever possible; graphical representations often provide a natural intermediate step between verbal and mathematical representations. If we underestimate the usefulness of graphical relationships, we risk struggling unnecessarily with the translation of verbal into mathematical descriptions.

P16. Use of functional relationships that are not interpretable We should be able to interpret functional relationships within the subject matter context of our problem; ecologically, physiologically, etc. Obviously, this applies only to those explanatory models for which such interpretability is an objective. If we cannot interpret some of the functional relationships, due to the connectedness of model components, we risk losing the ability to describe overall model behavior in terms of cause and effect.

Step II$_d$ – Estimate the parameters of the model equations

P17. Careless definition of dimensional units of model components We should take care in defining all model components precisely, including appropriate units of measure. If we are careless in our definitions, we risk making erroneous calculations due to dimensional incompatibility among model parameters, thus creating nonsensical variables which reduce model credibility.

P18. Use of coefficients without meaning to obtain dimensional consistency We should not create meaningless coefficients as a means of obtaining dimensional consistency among model components. The perceived need to create such coefficients results directly from careless definition of units of measure of model components (P17).

P19. Reluctance to use qualitative information We should not overlook the value of qualitative information in quantifying relationships within the model. If we try to rely solely on data to quantify our entire model, we risk a long and probably futile search for numerical information that does not exist. We almost always know more about the system-of-interest than we can confirm rigorously with data. We should not hesitate to create arbitrary scales of measurement (often unit-less indexes) to convert qualitative information to quantitative values; quantification does not imply accuracy, only precision. Once we have quantified a relationship, we can then determine much about the required numerical accuracy.

P20. Decision to remove a functional relationship due to lack of data We should resist the tendency to remove a functional relationship from the model because we can not find data with which it can be quantified. We almost always will commit a more serious error by removing an important relationship from the model than we will by guessing, and subsequently adjusting, its value.

P21. Reliance on "automated" model parameterization (fascination with methodology) We should resist the tendency to rely on automated ("canned" or "black box") methods to parameterize model relationships; we face a similar temptation during model evaluation (P30). On the one hand, we risk using an inappropriate methodology simply because it is convenient to do so and the methodology seems "almost right". On the other hand, we risk becoming overly fascinated with methodology per se, which can lead to a search for more elegant mathematical and statistical methodologies even though there is no objective demonstration they are needed.

P22. Use of overly sophisticated equations We should not use equations that are more sophisticated or complicated than necessary simply because they are more elegant. This is closely related to fascination with methodology (P21); we risk becoming distracted from our modeling objectives by a search for more elegant mathematical representations of system processes, even though there is no objective demonstration they are needed. An additional risk is the mathematical sophistication of an equation will imply an unwarranted confidence in our understanding of the relationship.

P23. Underestimation of the importance of negative feedback and time lags We should consider carefully our representation of negative feedback and time lags in our model equations. These are almost universal characteristics of complex systems, but are easily overlooked because cause and effect in complex systems are not always tightly linked in time and space. Common problems resulting from inappropriate representation of these characteristics include impossibly high or low values of system components and impossibly short response times.

Step II$_e$ – Execute baseline simulation
P24. Careless definition of baseline conditions We should take care in defining baseline system conditions, since these conditions often are used as the point of reference for both model evaluation and model application. Without well-conceived baseline conditions, we risk losing focus during model evaluation, particularly during sensitivity analysis. We also risk faulty experimental design during model application, leading to inappropriate analysis and interpretation of simulation results.

P25. Reliance on automated solutions to mathematical and programming problems We should not allow built-in programming safeguards to handle unforeseen inconsistencies in our mathematical formulation or programming. We risk aberrant model behavior resulting from inappropriate, automated procedures that avoid, rather than fix, the problem; we should provide the appropriate mathematical or programming solution ourselves. A common example is the attempt to divide by zero. Often the process being represented simply does not occur in nature when the system component in the denominator goes to zero; in this case, we could imbed the equation in an "if/then" statement that checks to see if the denominator is zero before attempting the division.

7.3 Phase III pitfalls: model evaluation

Step III$_a$ – Assess the reasonableness of the model structure and the interpretability of functional relationships within the model

P26. Underestimation of the importance of qualitative aspects of evaluation We should not underestimate the importance of qualitative aspects of model evaluation. We risk losing the most convincing defense of a dynamic model; the fact that all model components are reasonable and interpretable in ecological, or other appropriate subject-matter, terms.

P27. Acceptance of conceptually flawed functional relationships We should not accept functional relationships that we know are flawed, even if we have programmed the model (often via "if/then" statements) to prevent these functions from operating under those circumstances for which they are inadequate. Although we may have avoided the flaw numerically, we will have accepted as useful a model containing relationships that we know are conceptually flawed, thus obliging us to defend the model as a purely correlative model. Note avoiding the execution of flawed functional relationships is fundamentally different from avoiding the execution of mathematically undefined operations (P25).

P28. Acceptance or rejection of surprising model behavior without explanation We should not accept surprising model behavior until we have determined its cause; all model behavior is understandable because it is based on rules written in the form of mathematics and computer code. Surprising behavior can result from conceptual, logical, or computer coding errors, or can "emerge" from complex, but reasonable, interactions among model components. If we do not explain the source of surprising behavior, we risk doubts regarding model credibility.

Step III_b – Examine the correspondence between model behavior and the expected patterns of model behavior

P29. Inappropriate interpretation of the initial phase of model behavior We must take care to interpret appropriately the initial phase of model behavior; often our model will exhibit behavior patterns during the first several time steps that are fundamentally different from subsequent patterns. Differences between early and subsequent patterns usually result from differences between our estimates of the initial values of state variables and the values of state variables that are generated by, or "in step" with, functional relationships in the model. These differences may represent a "start-up" problem requiring re-initialization of the model, or, depending on model objectives, may represent precisely the behavior of most interest, as the system responds to the new circumstances that we want to explore.

P30. Reliance on "automated" model evaluation (fascination with methodology) We should resist the tendency to rely on automated ("canned" or "black box") methods to evaluate our model; we face a similar temptation during model parameterization (P21). We risk deceiving ourselves into thinking there is a rigorous mathematical and statistical protocol for model evaluation. We also risk, inadvertently, allowing our fascination with methodological details to distract our attention from a thorough evaluation of the real usefulness of the model.

Step III_c – Examine the correspondence between model projections and the data from the real system

P31. Tendency to believe all data from the real system are correct We should resist the tendency to believe data from the real system represent an absolutely reliable picture of real system dynamics; error-free data are extremely rare. We risk underestimating the usefulness of the model by interpreting all differences between model projections and real-system data as resulting from inadequacies in the model. Often these differences result from measurement errors, or other "noise", in the data. In fact, arguably, in some cases a simulated data point may be a better estimate than a data point from the real system.

P32. Inappropriate interpretation of statistical tests used in model evaluation We should not automatically assign importance to all statistically significant differences between model projections and real-system data. We risk underestimating the usefulness of the model based on differences that are of no practical significance; statistical and practical significance are not necessarily synonymous. We encounter an analogous pitfall during model application (P37).

Step III$_d$ – Determine the sensitivity of model projections to changes in the values of important parameters

P33. Careless design of sensitivity analysis We should design our sensitivity analysis to meet specific objectives, within the context of overall model objectives. We risk generating an enormous amount of useless information by attempting to conduct a "complete" sensitivity analysis involving many levels of each model parameter. We also risk generating misleading information by taking a "shotgun" approach to parameter selection, which can inadvertently bias our view of relative model sensitivity, and by varying parameter values over inappropriate ranges, which measure model response to situations that are impossible in the real system. We encounter an analogous pitfall during development of the experimental design for our model application (P35).

P34. Tendency to equate model sensitivity with model inadequacy We should resist the tendency to equate overall sensitivity of model behavior to changes in parameter values as a measure of model adequacy or inadequacy per se; we should interpret sensitivity analysis results on a parameter-specific basis within the context of model objectives. We risk underestimating the usefulness of the model as a tool for quantifying our uncertainty about real-system dynamics, and as a guide for prioritizing data needs. We also risk misinterpreting results of model application.

7.4 Phase IV pitfalls: model application

Step IV$_a$ – Develop and execute the experimental design for the simulations

P35. Careless experimental design for model application We should develop the experimental design for our model application carefully to meet specific project objectives; we must resist the tendency to simulate many scenarios just because we can do so easily. We risk generating useless, and potentially misleading, information by careless selection of the management and environmental scenarios we will simulate. We encounter an analogous pitfall during development of the design for our sensitivity analysis (P33).

P36. Tendency to overestimate the range of model applicability We must be careful to not apply the model to situations in which we violate the basic assumptions upon which the model is based. We risk generating meaningless or misleading results that may lead us astray in the development of subsequent models.

P37. Inappropriate interpretation of statistical tests used in model application We should not automatically assign importance to all statistically significant differences among simulated scenarios. We risk focusing attention on differences that are of no practical significance; statistical and practical significance are not necessarily synonymous. We encounter an analogous pitfall during model evaluation (P32).

Step IV$_b$ – Analyze and interpret the simulation results

P38. Failure to communicate numerical results directly in ecological terms We should take care to interpret all numerical results of our model application in ecological (or other appropriate subject-matter) terms whenever possible. We risk reducing the usefulness of our simulation results by describing them in a predominantly numerical context without a sufficient number of explicit links to the ecological context within which the numbers can be related directly to our initial questions.

Step IV$_c$ – Communicate the simulation results

P39. Tendency to imply false expectations regarding model projections We should take care to not lose sight of the fact that simulation results are the numerical consequences of following the logical and mathematical instructions which, collectively, are the model. We risk implying simulated data are not different from data based on direct physical measurements in the real world, and the model has permanence akin to that of a physical entity in the real world. We also risk implying that we can predict the future, which is theoretically impossible except for completely closed systems, which do not exist in the real world.

8

The modeling process in practice

In this chapter, we take a candid look at the practical application of simulation modeling. We suggest a strategy for model development that we have found helpful in traversing the pitfall-filled modeling landscape we depicted in Chapter 7. Although theoretically it is convenient to describe the modeling process as proceeding smoothly through the four phases described in Chapters 3 through 6, in practice we usually cycle through these phases several times. We seldom quantify the entire conceptual model before running simulations and evaluating model behavior. Rather, we usually construct a simple "running" model as quickly as possible and then expand it gradually through a series of small additions, which we carefully evaluate, until we have quantified the entire model (Fig. 8.1).

Thus, in practice, we engage ourselves in three general types of activities:

1 We develop a preliminary conceptual model of the entire system-of-interest, as well as a "plan of attack" for quantifying the model piece-by-piece.

2 We construct a series of intermediate developmental models, following our plan of attack, starting with what usually seems a trivially small portion of the overall model. We quantify, "run," and evaluate each model, and make any necessary adjustments, before proceeding to the next one. We add only a tiny piece to each new model, following our plan, which we also may modify as we proceed. We repeat the quantify, run, evaluate, and adjust sequence until we have reassembled the entire model, which seldom is identical to our original conceptual model.

3 We use the final, "reconstructed," model to address our questions, which we also may have modified as we sharpened our focus during model development.

This is an admittedly "ugly," seldom documented procedure, but each of the practical activities can be related directly to the theory described in Chapters 3 through 6. The iterative nature of the approach forces us to constantly reevaluate our model, both conceptually and quantitatively. This constant reevaluation sharpens our focus on project objectives, greatly reduces the likelihood the model code contains mathematical or logical errors, and invariably provides additional insight into the dynamics of the system-of-interest. The latter is, without a doubt, one of the most important and widely acknowledged benefits of the modeling process.

In the sections that follow, we relate the three activities that form our practical strategy for model development to the corresponding theoretical phases described in Chapters 3 through 6 (Fig. 8.1). Before proceeding, we suggest you take a moment to compare the details we present in Fig. 8.1 to the diagram we presented in Chapter 2 (Fig. 2.1) relating a common-sense approach to problem solving to modeling theory and modeling practice.

8.1 Preliminary conceptual model (*CM*)

The objective of the preliminary conceptual model is to qualitatively represent all relevant aspects of the system-of-interest as we currently understand them. This activity follows the first theoretical phase (I_a through I_f) relatively closely, however, we perform steps I_b through I_e simultaneously as we draw and redraw the box-and-arrow diagram that represents the system-of-interest (Fig. 8.1). Most commonly this activity is done with paper and pencil (and eraser!) rather than on a computer.

8.1.1 How to begin

We may have difficulty deciding how to begin developing the conceptual model. Theoretically, the system has no beginning or end, just connections among components, so we might start anywhere. However, a good rule of thumb is to begin thinking about the model in terms of the type of material that we want to represent and where that material accumulates. That is, we begin thinking about state variables. Often there is a natural flow of material passing through, or circulating within, the system that provides a "backbone" (a chain of state variables) for subsequent model structure. Energy flow through an ecosystem, with accumulation points of interest being plants, herbivores, and carnivores (Fig. 3.7), and nutrient cycling within an ecosystem, with accumulation points of

Theory		Pitfall	Practice				
Phase	Step	Number	*CM*	*IDM₁*	*IDMᵢ*	*IDMlast*	*FM*
I	a	1-3	✓	✓	...	✓	
	b	4, 5	✓	✓	...	✓	
	c	6	✓	✓	...	✓	
	d	7, 8	✓	✓	...	✓	
	e	9	✓	✓	...	✓	
	f	10	✓	✓	...	✓	
II	a	11, 12		✓	...	✓	
	b	13		✓	...	✓	
	c	14-16		✓	...	✓	
	d	17-23		✓	...	✓	
	e	24, 25		✓	...	✓	
III	a	26, 27		✓	...	✓	
	b	28, 29		✓	...	✓	
	c	30-32		✓	...	✓	
	d	33, 34				✓	
IV	a	35, 36					✓
	b	37					✓
	c	38, 39					✓

Fig. 8.1 Integrated view of ecological modeling, linking theory (Chapters 3–6), pitfalls (Chapter 7), and practice (this chapter). Checkmarks indicate the theoretical phases corresponding to the indicated stage of practical model development. Shaded areas represent theoretical phases that do not occur during the indicated practical stage.

Theoretical phases of model development. Phase I: Conceptual Model (Chapter 3, Fig. 3.1); Phase II: Quantitative Model (Chapter 4, Fig. 4.1); Phase III: Model Evaluation (Chapter 5, Fig. 5.1); Phase IV: Model Application (Chapter 6, Fig. 6.1).

Practical stages of model development. *CM*: conceptual model; *IDM*₁...ₗₐₛₜ: the series of intermediate developmental models; *IDM*ₗₐₛₜ: the last intermediate developmental model; *FM*: the final model.

interest being plants, herbivores, carnivores, and abiotic components (Fig. 3.8), are examples.

Having identified the state variables, this leads us quite naturally to think about the routes by which material moves into, out of, and among the state variables. That is, we begin thinking about material transfers. Identification of the material transfers, in turn, quite naturally focuses our attention on the information that is needed to control their rates. That is, we begin thinking about information transfers to each material transfer which arrive either directly or indirectly from other model components; state variables, driving variables, auxiliary variables, and/or constants. If we immediately think of some components that control a material transfer, we might add them to the model at the time we add the material transfer. But if we find ourselves pondering at length the various factors that may control any given material transfer, usually it is better to continue with the identification of state variables and material transfers, and postpone our contemplation of the more difficult aspects of model structure until we have identified all of the easier components.

8.1.2 Adding new components to the model

If we are in doubt with regard how to represent a new model component, a good rule of thumb is first to try the simplest possible representation: a constant. If it is not constant, try representing it as a driving variable. If it is affected by some other component in the system-of-interest, try representing it as an auxiliary variable. If the calculation of its new value requires knowledge of its previous value, represent it as a state variable, with, of course, the appropriate material transfers entering and/or leaving. Figures 3.2 through 3.5 demonstrate a series of alternative representations of available prey, in which we progressed from representation as an auxiliary variable (Figs 3.2–3.4) to representation as a state variable (Fig. 3.5).

8.1.3 Describing expected patterns

Once we have a conceptual model we think adequately represents the system-of-interest, we describe expected patterns of model behavior (I_f). This is also a paper-and-pencil activity that typically involves graphing expected temporal dynamics of key model components as well as graphically depicting our hypotheses concerning model predictions under different management policies or environmental conditions. These expectations may be based on data from direct observation or experimentation with the system-of-interest, theoretical relationships or generally applicable empirical relationships that are appropriate for the situation being modeled, qualitative information from the literature, or on opinions of experts.

8.1.4 Describing the plan of attack

Finally, based on the preliminary conceptual model, we outline a general plan for model development, that is, for quantifying the model. We first identify a tiny subset of the boxes and arrows in the preliminary conceptual model with which we will begin model quantification, then we identify a series of increasingly complex intermediate developmental models that we will quantify sequentially en route to obtaining the final

MODELING PRACTICE

model. Obviously, for extremely simple models we hardly need to elaborate a multi-step plan for quantitative model development. However, such a plan is useful for even relatively simple models, and for more complex models it is essential.

The objective of the series of intermediate developmental models is to obtain an adequate quantitative representation of the preliminary conceptual model. This activity includes all of the steps in the first (I_a through I_f), second (II_a through II_e) and third (III_a through III_d) theoretical phases, however, it departs from theory considerably in that we seldom quantify the entire conceptual model before running simulations and evaluating model behavior (Fig. 8.1). Rather, we usually construct a very simple "running" model as quickly as possible and then expand it through a series of tiny changes, following the general plan for model development we outlined when we finished the preliminary conceptual model (Section 8.1), until the model can be used to address our project objectives. We benefit greatly from having a running model as soon as possible because only then can we see the logical numerical consequences of our assumptions. Often the puzzle parts that fit together well conceptually do not fit together at all quantitatively. By making each addition to subsequent models as simple as possible, we not only facilitate the identification and correction of mathematical and logical errors, but also promote understanding of the relationships within the model that control system behavior.

As we proceed through the series of developmental models, our specific objectives (I_a), conceptualization of the system (I_b through I_e), and expected patterns of model behavior (I_f) likely will change from one intermediate model to the next, since we are shifting our focus from representation of one part of the system to another (Fig. 8.1). Likewise, the manner in which we quantitatively represent any given aspect of the system may change from one developmental model to the next, since the general strategy is to first greatly simplify and then gradually increase complexity. It is unlikely that we will change our choice of the general quantitative structure for the model (II_a), and, as we proceed, it becomes increasingly unlikely that we will change our choice of the basic time unit for simulations (II_b), but we quite likely will cycle through most of the other theoretical steps related to model quantification (II_c through II_e) and evaluation (III_a through III_c) for each new model (Fig. 8.1). We also may modify our general plan of attack for model development as we proceed, and may even decide to "fine tune" the objectives of the modeling project as we learn more about the system-of-interest from our interactions with the developing model.

We should proceed through the series of developmental models as quickly as possible. Obviously, "quickly" is a relative term, constrained

8.2 Intermediate developmental models (*IDM₍ᵢ₎*)

by the overall complexity of the model and the time we have to devote to the modeling project. But the goal is to maintain the continuity of our flow of ideas related to model development. Thus, even if we feel a strong need to refine our representation of a certain aspect of the model, we should postpone lengthy literature searches and time-consuming data analyses until we have the model fully quantified; that is, until we have reached the last intermediate developmental model (III$_d$) (Fig. 8.1). Most often we can obtain an adequate first approximation of the needed equations based on rough plots of available data or qualitative descriptions of functional relationships.

When we have fully quantified the last intermediate developmental model (IDM_{last}), we then focus on the necessary refinements to our representations of specific aspects of the model. Now is the time that we conduct literature searches and more detailed data analyses to confirm the functional forms of selected model equations and to improve the estimates of specific parameters. Having completed these final adjustments, we conduct our last, formal, evaluation of the model, which almost always includes a sensitivity analysis (we comment further on sensitivity analysis in Section 8.2.2). Assuming we fail to reject this last intermediate developmental model as useless, it becomes the final model that we will use to address the objectives of the modeling project.

8.2.1 Evaluate–adjust cycle for each developmental model

We now take a closer look at the manner in which we evaluate and adjust each of the developmental models before proceeding to the next. If we are dissatisfied with the current representation of any given developmental model because it fails to meet one of our first three evaluation criteria (III$_a$ through III$_c$), we return to an earlier theoretical step in model development to make appropriate modifications; the step to which we return depends on why we are dissatisfied (Fig. 8.2). (We usually do not conduct a sensitivity analysis (III$_d$) until we have reached the last developmental model.)

By far the most common type of adjustment we make during the entire modeling process involves quantitative "tuning" of the various developmental models; that is, the adjustment of functional forms of equations (II$_c$) and values of parameters (II$_d$). This is a legitimate activity during model development, but an activity that must be confined strictly in terms of (1) choosing parameters or functional forms to adjust, (2) limiting the number of parameters or functional forms adjusted, and (3) determining when to end "tuning" activities.

First, discretion must be used in choosing the parameters or functional forms to adjust as well as the types of adjustments to be made. Adjustments will be based on our best guess because no new information external to the model is being drawn on to make them, although the nature

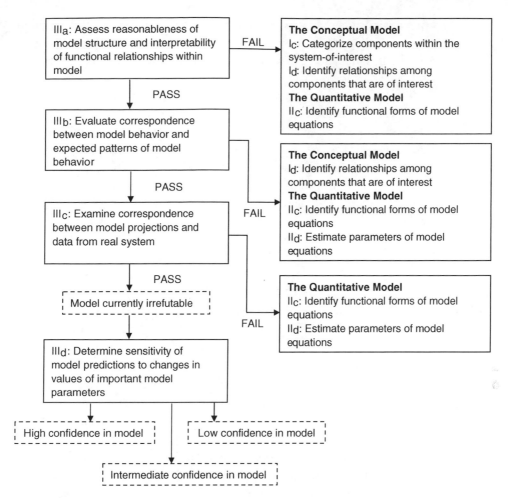

Fig. 8.2 Evaluate and adjust cycle for each developmental model. Sensitivity analysis is only conducted on the last intermediate developmental model.

of differences between model behavior and expected patterns of behavior may suggest specific alterations. But only those parameters or functional forms that we have specified in a tentative fashion should be altered, and the type of adjustment made must not result in a parameter value or functional form that can be refuted on conceptual grounds.

Second, we must limit the number of parameter values or functional forms of equations we adjust in any given model. There is no precise limit, but obviously, as the number of components adjusted increases, so does the number of possible combinations of adjusted values and, hence, the possibility that any improvement in model behavior will result solely from a fortuitous combination of adjustments. It is more common to adjust parameter values than functional forms of equations. This is understandable because the choice of functional forms usually has more profound implications concerning ecological interpretations of model structure than does the choice of parameter values. We are likely to have based our original choice of general functional forms on a relatively sound understanding of the nature of relationships in the real system, whereas we are less likely to have based our original choice of specific parameter values on an equally sound basis.

Finally, we should end all tuning of the model before proceeding with the formal evaluation (III_a through III_d) of the last developmental model (Figs 8.1, 8.2). If we make statistical comparisons between model projections and real-system data, this clearly marks the point at which tuning must end. If our comparisons between model projections and real-system data are nonstatistical, or we lack evaluation data from the real system, the line is perhaps less clear. But, in good faith, we must end tuning before we begin formal evaluation of the last developmental model.

8.2.2 Sensitivity analysis of last developmental model

This activity follows the corresponding theoretical phase (III_d) quite closely. Our objective is to determine the degree of response, or sensitivity, of model behavior to changes in various model components. We usually alter the value of one parameter at a time throughout an entire simulation and observe the subsequent effect on model behavior, but we may vary sets of parameters at the same time, or change the functional forms of relationships within the model, if these changes seem more appropriate. We usually try to vary any given parameter over a range of values indicative of the degree of uncertainty associated with our estimation of the parameter.

If we are lucky, model projections will be insensitive to changes in those parameters for which we have poor information. That is, the differences among model projections generated with the different parameter values will have no practical importance within the context of our objectives. If we are not so lucky, and differences in model projections are large enough to be of practical importance, all is not lost. In fact, this information is most useful in (1) identifying important gaps in our knowledge base, and (2) quantifying the uncertainty that we should attribute to our model projections. In such cases, a wise strategy is to make subsequent model projections, those we will use to address project objectives, with several versions of the model that, collectively, represent this uncertainty.

Thus, by indicating the degree to which uncertainty in our parameter estimates generates uncertainty in model projections, sensitivity analysis provides an appropriate context within which to make, and interpret, our projections with the final model (or the various versions thereof).

8.3 Final model (FM)

The objective of the final model is to address the objectives of the modeling project. This activity follows the last theoretical phase (IV_a through

IV$_c$) closely (Fig. 8.1). Because analysis of simulation results from our initial experimental design almost always raises new questions, before formally reporting our results, we commonly run additional simulations to explore new variants of the most interesting of the original scenarios. This is analogous to conducting a pilot study in the real world as a basis for future work.

Our simulated pilot study commonly consists of (1) an exploration of system behavior under relatively many variants of the selected scenarios and, perhaps, (2) a more formal analysis of some of the new scenarios. During exploratory simulations, we are interested primarily in identifying general trends in system behavior; if we have a stochastic model, we sometimes conduct these exploratory simulations using a deterministic version of the model in which all random variables are assigned their mean values. If some of the new scenarios seem worthy of more formal analysis, we develop and execute a new experimental design, analyze the new results, and incorporate these new findings into our final report.

We formally communicate results of our modeling project (IV$_c$), in general, just as we would results of field or laboratory projects. However, two items are worthy of note. First, since simulated experiments are much less subject to the constraints of time and money, it sometimes is difficult to identify a definitive ending point for the modeling project. Indeed, the modeling process ideally is a continuous one, and can help provide continuity in an on-going program of investigation. Nonetheless, we need to periodically report our results formally.

Second, the standard "methods" and "results" format of technical reports can be awkward for describing the development and application of a simulation model, since the model may be viewed as both a method and a result. We have found that replacing "methods" and "results," if possible, with headings such as "overview of the model," "model description," "model evaluation," and "simulation of [. . .]," provides a more comfortable format. We demonstrate use of this format in Appendix B, where we provide technical reports associated with the hunter-gatherer, population, and Commons examples.

9

The common–sense problems revisited

In this chapter, we relate the three activities that form our practical strategy for model development (Chapter 8, Fig. 8.1) to our common-sense solutions to the three problems presented in Section 2.1 (Fig. 2.1, Fig. 9.1). Recall that the parenthetical entries scattered through the three examples in Section 2.1, which are identical with those found in Figs 2.1 and 9.1, are intended to provide a useful cross-referencing of our common-sense approaches to the three problems with the steps involved in modeling theory and modeling practice. We suggest that you refer to these problems frequently as you read the remainder of this chapter with the dual goal of (1) appreciating the commonsensical nature of ecological modeling practice and (2) recognizing the added rigor provided by its theoretical basis.

9.1 Harvesting food for the winter (Section 2.1.1)

In our hunter-gatherer model, the material of interest was food items, their important points of accumulation were the forest and the total harvest, and the material transfer of interest moved food items from the forest (via harvest) to an accumulated total harvest (Figs 2.2a, 9.1). The rate of harvest depended on information about the current number of food items in the forest. Hence, we drew an information transfer from Forest Food Items (the state variable) to Harvest (the material transfer) (Figs 2.2b, 9.1).

9.1.1 The preliminary conceptual model (*CM*)

Regarding expected patterns of model behavior, we had expectations regarding the rate of decrease of food items in the forest during Novem-

Theory		Practice						
Phase	Step	CM	IDM$_i$				IDM$_{last}$	FM
			1	2	3	4		
I	a	H,P,C	P,C	P,C	C	C	H*,P,C	
	b	H,P,C	P,C	P,C	C	C	H*,P,C	
	c	H,P,C	P,C	P,C	C	C	H*,P,C	
	d	H,P,C	P,C	P,C	C	C	H*,P,C	
	e	H,P,C	P,C	C	C	C	H*,P,C	
	f	H,P,C	P,C	P,C	C	C	H*,P,C	
II	a		P,C	P*,C*	C*	C*	H,P*,C*	
	b		P,C	P*,C*	C*	C*	H,P*,C*	
	c		P,C	P,C	C	C	H,P,C	
	d		P,C	P,C	C	C	H,P,C	
	e		P,C	P,C	C	C	H,P,C	
III	a		P,C	P,C	C	C	H,P,C	
	b		P,C	P,C	C	C	H,P,C	
	c		P,C	P,C	C	C	H,P,C	
	d						H*,P*,C	
IV	a							H,P,C
	b							H,P,C
	c							H,P,C

Fig. 9.1 The integrated view of ecological modeling presented in Chapter 8 (Fig. 8.1) applied to the three examples from Chapter 2. H, P, C represent hunter-gatherer, population, and Commons examples, respectively. * indicates no change to the model was made during the indicated practical stage.

Theoretical phases of model development. Phase I: Conceptual Model (Ch. 3, Fig. 3.1), Phase II: Quantitative Model (Ch. 4, Fig. 4.1), Phase III: Model Evaluation (Ch. 5, Fig. 5.1), Phase IV: Model Application (Ch. 6, Fig. 6.1).

Practical stages of model development: *CM*: Conceptual Model; *IDM*$_{1...last}$: the series of intermediate developmental models; *IDM*$_{last}$: the last intermediate developmental model; *FM*: the final model.

ber (Figs 2.3, 9.1), and regarding the number of days it would take the hunter-gatherers to harvest the food they needed for winter (Section 2.1.1), based on expert opinion (of the hunter-gatherers). (We did not elaborate a multi-step plan for quantitative model development.)

9.1.2 The last (only) intermediate development model (*IDM$_{last}$*)

Our hunter-gatherer model was not complicated enough to warrant a series of intermediate developmental models. Our projections of system dynamics involved the calculation of just one material transfer, and the subsequent book-keeping associated with updating the values of the two state variables (*IDM*$_{last}$) (Section 2.1.1, Figs 2.2b, 9.1). Since simulated loss of food items from the forest and their accumulation in total harvest corresponded well with our expectations, we proceeded.

We used our final model (*FM*) (Figs 2.2b, 9.1) to project the decrease in the number of food items remaining in the forest and the resulting time taken to accumulate a total harvest of 75 food items, assuming that deforestation would result in only 80 items in the forest at the beginning of November (Fig. 2.3b). We suggested that the hunter-gatherers should have enough time to harvest the 75 items they need for winter. However, we noted that if winter storms arrived just a few days earlier than usual (than assumed in the model), the hunter-gatherers might not have enough time.

9.1.3 The final
model (*FM*)

In our population model, the material of interest was individuals, their important point of accumulation was in the population, and the material transfers of interest moved new individuals from a source (outside the system-of-interest) into the population (via births) and from the population into a sink (outside the system-of-interest) (via deaths) (Figs 2.4a, 9.1). Thus, by definition, we were not interested in where the new individuals came from or in what became of the dead individuals, only in the rates at which individuals were born (entered the system-of-interest from an infinite source) and died (left the system-of-interest and entered an infinite sink). The rate of births depended on information about the current number of individuals in the population and per capita birth rate, and the rate of deaths depended on information about the current number of individuals in the population and per capita death rate (Figs. 2.4b, 9.1). Per capita birth rate (an auxiliary variable) depended on the current number of individuals in the population (Figs 2.4c, 9.1), and per capita death rate (an auxiliary variable) depended on the occurrence of hurricanes (a driving variable) (Figs 2.4d, 9.1).

9.2 Estimating the probability of population extinction (Section 2.1.2)

9.2.1 The preliminary conceptual model (*CM*)

Regarding expected patterns of model behavior, we had expectations regarding dynamics of the population over the next 50 years (Figs 2.5a, 9.1) based on long-term observations of the population.

Regarding a plan of attack for developing the quantitative model, we first quantified a version of the model that represented both per capita birth rate and per capita death rate as constants, rather than as auxiliary variables (Fig. 2.4b). Next, we quantified a version of the model that still represented per capita death rate as a constant, but represented per capita birth rate as an auxiliary variable depending on population density (Fig. 2.4c). Finally, we quantified the full version of the model, which represented per capita death rate as an auxiliary variable depending on hurricanes (Fig. 2.4d).

9.2.2 The	We began with a simple intermediate developmental model that assumed
intermediate	no hurricanes, and constant birth and death rates (IDM_1) (Fig. 2.4b, 9.1).
development	Since we had removed the density-dependent, negative feedback of
models (IDM_i)	population size on birth rate from this version of the model, we
	expected J-shaped (exponential) rather than S-shaped (sigmoid) popula-
IDM_1	tion growth. Since simulated population growth was J-shaped, we
	proceeded.

IDM_2 We then replaced the density-dependent, negative feedback of population size on birth rate to form our second intermediate developmental model (IDM_2) (Figs 2.4c, 9.1). Since the density-dependent, negative feedback of population size on birth rate was once again included in the model, we once again expected the model should generate sigmoid growth (Figs 2.5a, 9.1). Since simulated population growth was S-shaped, we proceeded.

IDM_{last} We next replaced the effects of hurricanes on death rates to form our third, and last, intermediate developmental model (IDM_{last}) (Figs 2.4d, 9.1). Since simulated population levels in the presence of historical hurricane frequencies corresponded well with our expectations (Fig. 2.5), we proceeded.

9.2.3 The final
model (FM)

We used our final model (FM) (Figs 2.4d, 9.1) to project population dynamics and probabilities of extinction (Table 2.1) over the next 50 years, assuming annual hurricane probabilities of 0.10 and 0.50. We estimated probabilities of extinction within the next 50 years of <1 and 0.59 (Table 2.1b), respectively. Based on these results, we also decided to simulate intermediate hurricane probabilities to determine how much hurricane probabilities could increase before probability of population extinction would increase to dangerous levels. We suggested that as annual hurricane probabilities approached 40%, probability of extinction approached a dangerously high level (0.07, Table 2.1c).

9.3 Managing the Commons (Section 2.1.3)

9.3.1 The preliminary conceptual model (CM)

In our Commons model, we identified two materials of interest, forage biomass and animal weight units. Forage biomass accumulated on the Commons, and one material transfer of interest moved new forage biomass from a source into the Commons (via forage growth) (Figs 2.8a, 9.1), with the rate of forage growth depending on "per capita" forage growth rate (an auxiliary variable) and the current forage biomass on the Commons (Figs 2.8b, 9.1). Another material transfer of interest moved forage biomass from the Commons into a sink (via grazing loss), with the rate of grazing loss depending on current forage biomass on the Commons (Figs 2.8c, 9.1). Rate of grazing loss also depended on forage requirement of the herd of animals (an auxiliary variable), which, in turn, depended on forage requirement per animal (an auxiliary variable) and number of animals in the herd (a driving variable) (Figs 2.10a, 9.1). Another material transfer of interest moved forage biomass from the Commons into a sink (via nongrazing loss), with the rate of nongrazing loss depending on current forage biomass on the Commons (Fig. 2.8d). Rate of nongrazing loss also depended on nongrazing loss of the herd of animals, which, in turn, depended on nongrazing loss rate per animal (a constant) and number of animals in the herd (a driving variable) (Figs 2.10a, 9.1).

Animal weight units accumulated within a representative individual animal, and one material transfer of interest was weight gain (Figs 2.9a, 9.1). Rate of weight gain depended on the current weight of the animal and the maximum possible "per capita" rate of weight gain (an auxiliary variable), which also depended on the current weight of the animal (Figs 2.9b, 9.1). Rate of weight gain also depended on a weight gain index (an auxiliary variable), which, in turn, depended on relative forage availability (an auxiliary variable) (Figs 2.10b, 9.1). Another material transfer of interest was weight loss which depended on current weight of the animal and maintenance requirements (a constant) (Figs 2.9c, 9.1).

Regarding expected patterns of model behavior, we had expectations regarding effects of grazing on annual growth of forage biomass (Figs 2.7, 9.1), based on experimental data from a similar system, regarding long-term fluctuations in forage biomass (Figs 2.11, 9.1), based on long-term observations of forage biomass on the Commons, and regarding the manner in which individual animals should gain weight (Figs 2.11, 9.1) based on observations of initial and final weights of animals that had been on the Commons.

Regarding a plan of attack for developing the quantitative model, we first quantified a version of the model that represented forage growth without grazing or nongrazing losses (Fig. 2.8b). Next, we planned to add grazing losses (Fig. 2.8c) and then nongrazing losses (Fig.2.8d) to the model, and then proceed to focus on the animal components of the model (Fig. 2.9). However, we decided to change our plan of attack and leave the addition of grazing and nongrazing losses until after we had quantified the basic dynamics of the grazing animals. Thus, we proceeded to quantify a version of the model that simulated animal weight gain, without including weight loss due to maintenance requirements (Fig. 2.9b). We then added weight losses (Fig. 2.9c). We then coupled the forage and animal submodels by first representing the effect of grazing and trampling (nongrazing loss) on forage dynamics (Fig. 2.10a), then by adding the effects of forage dynamics on animal weight dynamics (Fig. 2.10b).

We first focused on simulating forage growth, without grazing or non-grazing losses (IDM_1) (Figs 2.8b, 9.1). Since the simulation results from IDM_1 corresponded well with results from the "no-grazing" treatment of the field experiment to which we had access (Figs 2.13, 9.1), we proceeded.

At this point, we decided to modify our general plan of attack for model development. Rather than next adding grazing losses and then nongrazing losses to the model, as we initially had planned, we decided to quantify a version of the model that simulated animal weight gain, without including weight loss due to maintenance requirements (IDM_2) (Figs 2.9b, 9.1). When we compared simulated weight gain to our expectations, we noticed our simulated growth curve was S-shaped (sigmoid) while we had expected a linear growth curve (Figs 2.15, 9.1). Thus, we were obliged to reconcile these differences. After reflection, we decided our initial expectation was flawed, so we proceeded.

We next added weight loss due to metabolism (IDM_3) (Figs 2.9c, 9.1). As we were quantifying IDM_3, we decided not to draw upon some of the information we had compiled, that dealing with the energetic costs of reproduction, because it was irrelevant within the time frame we were simulating (1 year). Since our simulated growth curve was S-shaped and lower than that produced without metabolic losses (Fig. 2.16), as expected, we proceeded.

We proceeded to couple the two submodels by representing the effect of grazing and trampling on forage dynamics (IDM_4) (Figs 2.10a, 9.1), still assuming forage dynamics were not affecting animal weight dynamics. Since the simulation results from IDM_4 corresponded reasonably well with results from the two grazing treatments of the field experiment to which we had access (Figs 2.17), we proceeded to add the effect of forage dynamics on animal weight dynamics (IDM_{last}) (Figs 2.10b, 9.1). We again compared simulation results to results of the two grazing experiments (Fig. 2.19), and also to our expectations of long-term forage dynamics (Fig. 2.20a) and of final animal weights (Fig. 2.20b). We were satisfied with the performance of IDM_{last}, but we remained a bit concerned about the effect of the uncertainty with which we had estimated a particularly important parameter (trampling rate) on our ability to project the various grazing (and trampling) scenarios of interest.

Thus, we conducted a sensitivity analysis on IDM_{last} to determine how sensitive our projections of both forage and animal weight dynamics were to changes in our estimation of trampling rate (nongrazing loss). Since results of sensitivity analysis indicated our projections at high animal stocking rates differed noticeably depending on our estimate of trampling rate (Figs 2.22, 9.1), we decided to proceed with three versions of our model, each with a different estimate of trampling rate.

We used the three versions of our final model (*FM*) (Figs 2.10b, 9.1) to project forage and animal weight dynamics (Figs 2.23–2.26, 9.1) over the next 10 years, assuming we (1) did and (2) did not keep up with our neighbor's planned increases in stocking rate. Our projections indicated that both forage biomass and individual animal weights at harvest would decrease under either of the two stocking scenarios. Based on these results, we also explored an "early harvest" scenario (Fig. 2.27) that occurred to us after considering the results obtained from our initial experimental design. We suggested that early harvest may be a viable option in terms of sustaining forage resources, however, it does not alleviate the problem of dangerously low individual animal weights at harvest. Since projections from all three versions of our FM yielded the same general trends (Figs 2.23–2.27), we could answer our initial questions with relatively high confidence, in spite of the uncertainty in our estimates of trampling rate. If projections from the different versions had yielded different answers to our questions, we would have concluded that, based on current information, we simply can not answer our questions with confidence.

9.3.3 The final model (*FM*)

10

Reflections

In closing, we would like to offer some reflections on how ecological modeling fits into the broader context of problem solving. More specifically, we would like to comment on the systems approach as a complement to other methods of problem solving, on ecological modeling as a problem-solving process, and on the expectations we should (and should not) have of the ecological models that are the product of that process.

10.1 The systems approach as a complement to other methods of problem solving

The systems approach certainly is not the only useful approach to problem solving. Throughout the history of mankind, trial and error has been by far the most widespread and useful method. Unfortunately, for some problems, the appropriate trials are too long and the possible errors are too costly. The scientific method of solving problems emphasizes more disciplined observation and perhaps manipulation of particularly interesting parts of the world. Scientists formally interpret these observations in a variety of qualitative (description and classification) and quantitative (mathematical and statistical analyses) ways, depending on the type of problem, or type of system, with which they are dealing. Scientists and other problem solvers use the systems approach to integrate relevant information gained from trial and error or expert opinion and from the scientific method in a form that facilitates formal description of the structure and dynamics of complex systems.

Precise definition of a "complex system" is neither possible nor necessary. However, it is possible and useful to relate types of systems to formal methods of problem solving in a very general way. For example, we might characterize systems in terms of number of components and degree of interrelatedness of components (Fig. 10.1). Problems related to systems with relatively few, highly interrelated components can be addressed in analytical form mathematically. Physicists deal with mechanical systems in this manner – Newton's laws of motion are an example. Problems related to systems with relatively many, loosely related components can be addressed statistically. For example, the movement of gas molecules inside a closed, expandable container can be viewed as random and interesting system relationships can be described in terms of average temperature, pressure, and volume. But problems related to systems with relatively many, closely interrelated components cannot be addressed effectively by either of these two problem-solving methods. On the one hand, such systems usually cannot be solved mathematically because an analytical solution to the set of equations describing the system does not exist. On the other hand, dynamics of these systems cannot be represented statistically as average tendencies because interrelatedness of components, or system structure, causes markedly nonrandom behavior. Systems analysis and simulation focus specifically on these "intermediate" systems characterized by "organized complexity" in which system structure both controls and is changed by system dynamics.

From a slightly different perspective, we might compare methods of problem solving in terms of the relative level of understanding and the amount of data available on the system in which we are interested (Fig. 10.2). As used here, understanding refers to the informal integration of

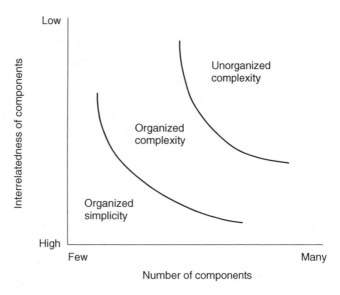

Fig. 10.1 General characterization of different types of systems in terms of the number of components and the degree of interrelatedness of components (modified from Weinberg, 1975).

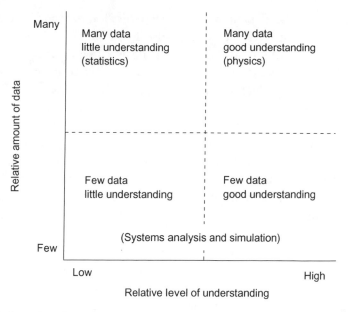

Fig. 10.2 Comparison of methods of problem solving in terms of the relative level of understanding and the relative amount of data available on the system (modified from Holling, 1978, and Starfield and Bleloch, 1986).

all that we have learned about the system by any means. The amount of data refers to the degree to which we have good data on all relevant aspects of the system. If we understand the structure and general dynamics of the system and have good data on all important processes occurring within the system, we often can develop mathematical models and solve them analytically. If we have good data but lack understanding of the underlying system processes that generated them, we often can use statistical analyses to search for patterns in the data that will help us hypothesize the nature of these underlying processes. If we have relatively few data but at least some understanding of system structure and general dynamics, we often can use systems analysis and simulation to investigate our hypotheses about how the system works. Obviously, if we cannot formulate useful hypotheses concerning system structure and function based on our current understanding, we should focus our efforts on further observation of the system.

Of course, in practice, domains of the various methods of problem solving overlap broadly. The most useful method for a given problem at a particular time depends on our conceptualization of the problem itself, which places us in one of the regions in Fig. 10.1, and on the current state of knowledge about the problem within this conceptual framework, which places us in one of the regions of Fig. 10.2. Relative usefulness of the various methods also changes as we continue to work on a given problem, that is, as we increase understanding and accumulate data. We might imagine a scenario in which we start with few data and little understanding. We first integrate existing knowledge via systems analysis and the use of simulation models to generate hypotheses about how the system works. This model-building process increases our understanding of the system and identifies specific areas in which important data are lacking. As we accumulate more relevant data, we use statistical analyses to interpret these data, thus generating further hypotheses concerning system structure and function that can be incorporated into our simulation models. Further simulation increases understanding and identifies more data needs, and so on. Theoretically, as we continue to accumulate relevant data and increase our knowledge of the system, we eventually achieve complete understanding of the system and the definitive solution of our problem.

As ecological modelers, or ecologists and natural resource managers who develop and use ecological models, we deal primarily with systems characterized by "organized complexity" for which we have relatively few data and little hope of ever accumulating a "complete" data set. That is, we find ourselves dealing with exactly those sorts of systems for which systems analysis and simulation were developed. The systems approach does not replace other methods of problem solving but, rather, provides a framework that allows effective integration of knowledge gained from description, classification, and mathematical and statistical analysis of our observations of the world.

10.2 Ecological modeling as a problem-solving process

We began Chapter 1 by stating that the usefulness of ecological modeling results as much from the process (problem specification, model development, and model evaluation) as from the product (the final model and simulations of system dynamics). One of the greatest benefits of the modeling process within the contexts of ecological research, natural resource management, and environmental education is that it facilitates unambiguous communication among multidisciplinary, multivocational groups of problem solvers. This benefit is well known (e.g., see Holling, 1978; Grant, 1998; van den Belt, 2004), but arguably we do not take advantage of this benefit as often as we should.

Within the context of ecological research, the modeling process facilitates the integration of multidisciplinary teams of researchers and enhances research continuity by recording, in a dynamic way, via the series of intermediate developmental models, the history of the program. As scientists and graduate students join and leave the program, and as perceptions of individual investigators change, the evolving model provides a common point of reference for incorporation of new members and the context within which past, present, and proposed projects can be integrated and evaluated.

THEORY, PRACTICE, AND COMMON SENSE

Within the context of natural resource management, the modeling process provides a basis for sustainable, multiple use of natural resources by facilitating multidisciplinary planning and creation of an effective communication interface between scientists and policy makers. Practical application of the process often involves a series of workshops in which a team of disciplinary specialists, systems modelers, and decision makers focuses on development of a simulation model to address specific management questions. During the series of workshops, the simulation model evolves into the communication interface between scientists and decision makers as team members develop a sense of joint ownership of the model. The joint participation of scientists and policy makers in the entire modeling process often does more to foster truly adaptive management of natural resources than does the detailed analysis of particular simulation results produced by the model.

Within the context of environmental education, the modeling process facilitates integrative thinking, active versus passive learning, and effective communication. The systems perspective provides the common conceptual framework and vocabulary needed to develop integrated educational programs that promote interdisciplinary communication and active intellectual teamwork within a problem-solving environment. The process of developing a quantitative systems model requires precise representation of ideas concerning relationships controlling system dynamics. Whereas ordinary spoken and written language allows ambiguous, incomplete, and even illogical statements to go unnoticed, systems simulation demonstrates the numerical consequences of assumptions and logic and identifies flaws in the understanding of causal relationships. The opportunity for modelers to investigate rigorously and rapidly the effects on system dynamics of altering assumptions and logic fosters active participation in the learning process.

So, if the usefulness of ecological modeling results as much from the process as from the product, what expectations should we have for the final model? That is, what should we expect to learn from the simulation results we generate using our final model?

10.3 Expectations for ecological models

One expectation that we definitely should not have is that we will be able to predict (with absolute certainty) the future. Perhaps this seems a silly thing to mention. But the tendency to implicitly append "with absolute certainty" to "predict" is perhaps the most widespread and dangerous misconception about models, ecological or otherwise. Although we commonly phrase our model objectives in terms of "predicting" the future behavior of our system-of-interest under different scenarios, what we really mean is that we will project, always with some degree of uncertainty, possible alternative futures under certain restrictive assumptions.

Semantics aside, the point is that predicting/projecting/forecasting/ prophesying the future with complete certainty is not a legitimate expectation of our final model.

However, we should expect that simulation results we generate with our final model will provide new knowledge that will help us reduce, in a useful way, the uncertainty with which we view the future of our system-of-interest. That is, we should expect to gain knowledge from our simulations that will allow us to discard as highly unlikely some system futures that previously seemed quite plausible. The relative amount of knowledge we gain depends in large part on the current level of knowledge about the system-of-interest. Logically, the less we currently understand about the system, the more we are likely to learn. The fact that we are using the systems approach and simulation to solve our problem implies that we are dealing with a system for which we have relatively few data and, quite likely, a relatively low level of understanding (Fig. 10.2).

Related to the legitimate expectation that we will gain useful knowledge about our system-of-interest from our final model is the erroneous expectation that the amount of knowledge we gain will be directly related to model complexity. The amount of knowledge we gain is, indeed, related to model complexity, but the form of the relationship depends on our current level of knowledge about the system (Fig. 10.3). As we

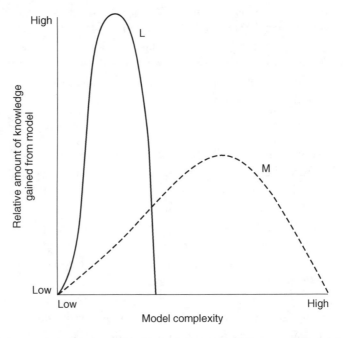

Fig. 10.3 General relationships between model complexity and the relative amount of knowledge gained from the model in systems for which we currently have much (M) and little (L) knowledge (modified from Jørgensen and Bendoricchio, 2001).

increase model complexity in systems for which we have little knowledge, the relative amount of knowledge gained will increase rapidly at first, and relative gains will be great. Even the very general patterns of behavior generated by simple models will add substantively to our knowledge. But soon further increases in complexity will yield little, and finally no, new knowledge. The benefits to model performance associated with explicitly representing more system components will be outweighed by the detriments associated with the cumulative effect of errors in parameter estimation. As we increase model complexity in systems for which we have much knowledge, the relative amount of knowledge gained will increase slowly at first, and relative gains will be small. Most of the general patterns of behavior generated by simple models already will be a part of our knowledge base. However, as complexity continues to increase, the relative amount of knowledge gained will begin to increase more rapidly. Eventually, the benefits to model performance will be outweighed by the detriments, just as in systems for which we have little knowledge, but at a higher level of model complexity. By definition, the relative amount of knowledge gained will be less in systems for which we already have much knowledge.

10.4 A final thought

We began this book by stating that the usefulness of ecological simulation modeling results as much from the modeling process as from the resulting models, and that skill in the process is gained through practice. We hope this book has provided not only what you need to know to begin building and using ecological models in a useful manner, but also the desire and courage to do so. Regardless of your current level of ecological, mathematical, and computer programming expertise, we believe you can apply the modeling process as described in this book to learn more about dynamic ecological systems. In the process, as often happens, you are quite likely to pick up a few more mathematical and programming skills, and perhaps may even team up with a mathematician or a programmer. In any event, your modeling experience will serve you well in identifying the information and expertise that you need to address your questions adequately, and will facilitate your communication with those who can provide that information and expertise.

References

Cochran, W.G. and G.M. Cox. 1957. *Experimental Design*. New York: John Wiley.

Forrester, J.W. 1961. *Industrial Dynamics*. Cambridge, MA: MIT Press.

Gilman, M. and R. Hails. 1997. *An Introduction to Ecological Modelling: putting theory into practice*. Oxford: Blackwell Science.

Grant, W.E. 1986. *Systems Analysis and Simulation in Wildlife and Fisheries Sciences*. New York: John Wiley.

Grant, W.E. 1998. Ecology and natural resource management: reflections from a systems perspective. *Ecological Modelling* 108:67–76.

Grant, W.E., E.K. Pedersen, and S.L. Marín. 1997. *Ecology and Natural Resource Management: systems analysis and simulation*. New York: Wiley.

Hardin, G. 1968. The tragedy of the commons. *Science* 162:1243–1248.

Holling, C.S. 1978. *Adaptive Environmental Assessment and Management*. New York: John Wiley.

Innis, G.S. 1979. A spiral approach to ecosystem simulation, I. In G.S. Innis and R.V. O'Neill (eds.), *Systems Analysis of Ecosystems*. Burtonsville, MD: International Cooperative Publishing House, pp. 211–386.

Jørgensen, S.E. and G. Bendoricchio. 2001. *Fundamentals of Ecological Modelling*, 3rd edn. Oxford: Elsevier Science.

Ott, L. and M. Longnecker. 2001. *An Introduction to Statistical Methods and Data Analysis*, 5th edn. Pacific Grove, CA: Duxbury Press.

Rykiel, E.J., Jr. 1996. Testing ecological models: the meaning of validation. *Ecological Modelling* 90:229–244.

Smith, F.E. 1973. Analysis of ecosystems. In D.E. Reichle (ed.), *Analysis of Temperate Forest Ecosystems*. New York: Springer-Verlag.

Spain, J.D. 1982. *BASIC Microcomputer Models in Biology*. Reading, MA: Addison-Wesley.

Starfield, A.M. and A.L. Bleloch. 1986. *Building Models for Conservation and Wildlife Management*. New York: Macmillan.

Steinhorst, R.K. 1979. Parameter identifiability, validation, and sensitivity analysis of large system models. In G.S. Innis and R.V. O'Neill (eds.), *Systems Analysis of Ecosystems*. Burtonsville, MD: International Cooperative Publishing House.

van den Belt, M. 2004. *Mediated Modeling: a systems dynamics approach to environmental consensus building.* Washington: Island Press.

van Dyne, G.M. 1969. *The Ecosystem Concept in Natural Resource Management.* New York: Academic Press.

Weinberg, G.M. 1975. *An Introduction to General Systems Thinking.* New York: John Wiley.

Appendix A

Introduction to the ecological modeling literature

Complexity is a hallmark feature of all ecological systems and simulation modeling is one method (one of the best!) to approach ecological/environmental problems. The field of ecological modeling is incredibly diverse and simulation has been applied to every aspect of ecology. In this text, we built simulation models for three relatively simple examples to emphasize the modeling process in a commonsensical, straightforward way. As we have mentioned before, the process is as important as the final model, however, rarely is the entire process documented (van den Belt (2004) has dedicated a text to describing the entire modeling process for several case studies). In the peer-reviewed literature, it is not practical to publish each developmental model and most often, we are only presented with the final model. Attempting to provide a review of published models is a daunting, almost impossible, task because the literature is replete with examples covering the entire spectrum of ecological thought.

We recognize that the problems an ecological modeler will be asked to solve are more complex than our examples, and we feel that providing references to some of the models in the peer-reviewed literature is important. Below we provide references for sixty simulation models. These examples will hopefully provide newcomers to the field with a starting place to explore the literature and provide experienced modelers with more references. We are not endorsing these models as better or worse than others, in fact, by providing references to only a few, we realize we are leaving out excellent models, and this is not our intention. In fact, we chose these models by using online search engines for the major ecological journals using the key words "simulation model." We divided the articles into four sections: ecosystem models, community models, population models, and applied models. All of these articles emphasize the intricacies involved in modeling complex ecological systems. Ecological modeling takes practice, and now that you have read through the text and are familiar with the simulation modeling process, we invite you to peruse these articles (and literature cited sections) to find ideas for your own ecological models. Good luck!

Ecosystem models   Botkin, D.B., J.F. Janak, and J.R. Wallis. 1972. Some ecological consequences of a computer model of forest growth. *Journal of Ecology* 60:849–872.

DeAngelis, D.L., L.J. Gross, M.A. Huston, et al. 1998. Landscape modeling for Everglades ecosystem restoration. *Ecosystems* 1:64–75.

Gbondo-Tugbawa, S.S., C.T. Driscoll, M.J. Mitchell, J.D. Aber, and G.E. Likens. 2002. A model to simulate the response of a northern hardwood forest ecosystem to changes in S deposition. *Ecological Applications* 12:8–23.

Grunwald, K.R. Reddy, J.P. Prenger, and M.M. Fisher. 2007. Modeling of the spatial variability of biogeochemical soil properties in a freshwater ecosystem *Ecological Modelling* 201:521–535S.

Hall, G.M.J. and D.Y. Hollinger. 2000. Simulating New Zealand forest dynamics with a generalized temperate forest gap model. *Ecological Applications* 10:115–130.

Kadlec, R.H. and D.E. Hammer. 1988. Modeling nutrient behavior in Wetlands. *Ecological Modelling* 40:37–66.

Lauenroth, W.K., D.L. Urban, D.P. Coffin, W.J., et al. 1993. Modeling vegetation structure–ecosystem process interactions across sites and ecosystems. *Ecological Modelling* 67:49–80.

McKane, R.B., E.B. Rastetter, G.R. Shaver, K.J., et al. 1997. Climatic effects on tundra carbon storage inferred from experimental data and a model. *Ecology:* 78:1170–1187.

Montoya, R.A., A.L. Lawrence, W.E. Grant, and M. Velasco. 2000. Simulation of phosphorus dynamics in an intensive shrimp culture system: effects of feed formulations and feeding strategies. *Ecological Modelling* 129:131–142.

Nonaka, E. and T.A. Spies. 2005. Historical range of variability in landscape structure: a simulation study in Oregon, USA. *Ecological Applications* 15:1727–1746.

Rivera, E.C., J. Ferraz de Queiroz, J.M. Ferraz, and E. Ortega. 2007. Systems models to evaluate eutrophication in the Broa Reservoir, São Carlos, Brazil. *Ecological Modelling* 202:518–526.

van der Peijl, M.J., M.M.P. van Oorschot, and J.T.A. Verhoeven. 2000. Simulation of the effects of nutrient enrichment on nutrient and carbon dynamics in a river marginal wetland. *Ecological Modelling* 134:169–184.

Walker, N.A., H.A.L. Henry, D.J. Wilson, and R.L. Jefferies. 2003. The dynamics of nitrogen movement in an Arctic salt marsh in response to goose herbivory: a parameterized model with alternate stable states. *Journal of Ecology* 91:637–650.

Walters C.J. and I.E. Efford. 1972. Systems analysis in the Marion Lake IBP Project. *Oikos* 11:33–44.

Williamson, M.S., T.M. Lenton, J.G. Shepherd, and N.R. Edwards. 2006. An efficient numerical terrestrial scheme (ENTS) for Earth system modelling. *Ecological Modelling* 15:362–374.

Community models   Alroy, J. 2001. A multispecies overkill simulation of the end-Pleistocene megafaunal mass extinction. Science 292:1893–1896.

Blackwell, G.L., M.A. Potter, and E.O. Minot. 2001. Rodent and predator population dynamics in an eruptive system. *Ecological Modelling* 142:227–245.

Bleher, B., R. Oberrath, and K. Böhning-Gaese. 2002. Seed dispersal, breeding system, tree density and the spatial pattern of trees – a simulation approach. *Basic and Applied Ecology* 3:115–123.

DeAngelis, D.L. and J.H. Petersen. 2001. Importance of the predator's ecological neighborhood in modeling predation on migrating prey. *Oikos* 94:315–325.

Firle, S., R. Bommarco, B. Ekbom, and M. Natiello. 1998. The influence of movement and resting behavior on the range of three carabid beetles. *Ecology* 79:2113–2122.

Gertseva, V.V., J.E. Schindler, V.I. Gertsev, N.Y. Ponomarev, and W.R. English. 2004. A simulation model of the dynamics of aquatic macroinvertebrate communities. *Ecological Modelling* 176:173–186.

Jørgensen, S.E. and B.D. Fath. 2004. Modelling the selective adaptation of Darwin's Finches. *Ecological Modelling* 176:409–418.

Jørgensen, S.E., E. Verdonschot, and S. Lek. 2002. Explanation of the observed structure of functional feeding groups of aquatic macro-invertebrates by an ecological model and the maximum exergy principle. *Ecological Modelling* 158:223–231.

Liedloff, A.C. and G.D. Cook. 2007. Modelling the effects of rainfall variability and fire on tree populations in an Australian tropical savanna with the Flames simulation model *Ecological Modelling* 201:269–282.

Risser, P.G. and J.B. Mankin. 1986. Simplified simulation model of the plant producer function in shortgrass steppe. *American Midland Naturalist* 115:348–360.

Seabloom, E.W. and O.J. Reichman. 2001. Simulation models of the interactions between herbivore foraging strategies, social behavior, and plant community dynamics. *American Naturalist* 157:76–96.

Takenaka, A. 2006. Dynamics of seedling populations and tree species coexistence in a forest: a simulation study. *Ecological Research* 21:356–363.

Watkinson, A.R. and J.C. Powell. 1993. Seedling recruitment and the maintenance of clonal diversity in plant populations – a computer simulation of *Ranunculus repens*. *Journal of Ecology* 81:707–717.

Wiegand, K., K. Henle, and S.D. Sarre. 2002. Extinction and spatial structure in simulation models. *Conservation Biology* 16:117–128.

Wu, J., J.L. Vankat, and Y. Barlas. 1993. Effects of patch connectivity and arrangement on animal metapopulation dynamics: a simulation study. *Ecological Modelling* 65:221–254.

Abramsky, Z. and G.M. van Dyne. 1980. Field studies and a simulation model of *Population models* small mammals inhabiting a patchy environment. *Oikos* 35:80–92.

Brisson, J. and J.F. Reynolds. 1997. Effects of compensatory growth on population processes: a simulation study. *Ecology* 78:2378–2384.

Essington, T.E. 2003. Development and sensitivity analysis of bioenergetics models for skipjack tuna and albacore: a comparison of alternative life histories. *Transactions of the American Fisheries Society* 132:759–770.

Fong, P. and P.W. Glynn. 2000. A regional model to predict coral population dynamics in response to El Niño–southern oscillation. *Ecological Applications* 10:842–854.

Gerber, L.R., M.T. Tinker, D.F. Doak, J.A. Estes, and D.A. Jessup. 2004. Mortality sensitivity in life-stage simulation analysis: a case study of southern sea otters. *Ecological Applications* 14:1554–1565.

Haas, H.L., K.A. Rose, B. Fry, T.J. Minello, and L.P. Rozas. 2004. Brown shrimp on the edge: linking habitat to survival using an individual-based simulation model. *Ecological Applications* 14:1232–1247.

Haight, R.G., D.J. Mladenoff, and A.P. Wydeven. 1998. Modeling disjunct gray wolf populations in semi-wild landscapes. *Conservation Biology* 12:879–888.

Herban, T., H. Rydin, and L. Soderstrorm. 1991. Spore establishment probability and the persistence of the fugitive invading moss, *Orthodontium lineare*: a spatial simulation model. *Oikos* 60:215–221.

Liljesthröm, G. and J. Rabinovich. 2004. Modeling biological control: the population regulation of *Nezara viridula* by *Trichopoda giacomellii*. *Ecological Applications* 14:254–267.

Rodenhouse, N.L., T.W. Sherry, and R.T. Holmes. 1997: Site-dependent regulation of population size: a new synthesis. *Ecology* 78:2025–2042.

Rose, K.A., E.S. Rutherford, D.S. McDermot, J.L. Forney, and E.L. Mills. 1999. Individual-based model of yellow perch and walleye populations in Oneida Lake. *Ecological Monographs* 69:127–154.

Scanlan, J.C., D.M. Berman, and W.E. Grant. 2006. Population dynamics of the European rabbit (*Oryctolagus cuniculus*) in north eastern Australia: simulated responses to control. *Ecological Modelling* 196:221–236.

Sharov, A.A. and J.J. Colbert. 1996. A model for testing hypotheses of gypsy moth, *Lymantria dispar* L., population dynamics. *Ecological Modelling* 84:31–51.

Stenseth, N.C. 1980. Modelling the population dynamics of voles: models as research tools *Oikos* 29:449–456.

Tixier, P., J.-M. Risède, M. Dorel, and E. Malézieux. 2006. Modelling population dynamics of banana plant-parasitic nematodes: a contribution to the design of sustainable cropping systems. *Ecological Modelling* 198:321–331.

Applied models   Cheng, J.A. and J. Holt. 1990. A systems analysis approach to brown planthopper control on rice in Zhejiang province, China: a simulation of outbreaks. *Journal of Applied Ecology* 27:85–99.

Grant, W.E., W.T. Hamilton, and E. Quintanilla. 1999. Sustainability of agroecosystems in semi-arid grasslands: simulated management of woody vegetation in the Rio Grande Plains of southern Texas and northeastern Mexico. *Ecological Modelling* 124:29–42.

Grevstad, F.S. 1999. Factors influencing the chance of population establishment: implications for release strategies in biocontrol. *Ecological Applications*. 9:1439–1447.

Hart, R.A., J.W. Grier, and A.C. Miller. 2004. Simulation models of harvested and zebra mussel colonized three ridge mussel populations in the upper Mississippi River. *American Midland Naturalist* 151:301–317.

Hels, T. and G. Nachman. 2002. Simulating viability of a spadefoot toad *Pelobates fuscus* metapopulation in a landscape fragmented by a road. *Ecography* 25:730–744.

Liu, J. and P.S. Ashton. 1999. Simulating effects of landscape context and timber harvest on tree species diversity. *Ecological Applications* 9:186–201.

Malanson, G.P., Q. Wang, and J.A. Kupfer. 2007. Ecological processes and spatial patterns before, during and after simulated deforestation *Ecological Modelling* 202:397–409.

Marín, S.L., R. Westermeier, and J. Melipillán. 2002. Simulation of alternative management strategies for red algae, luga roja (*Gigartina skottsbergii* Setchell and Gardner) in southern Chile *Ecological Modelling* 154:121–133.

Pedersen, E.K., J.W. Connelly, J.R. Hendrickson, and W.E. Grant. 2003. Effect of sheep grazing and fire on sage grouse populations in southeastern Idaho. *Ecological Modelling* 165:23–47.

Phillips, P.D., C.P. de Azevedo, B. Degen, I.S. Thompson, J.N.M. Silva, and P.R. van Gardingen. 2004. An individual-based spatially explicit simulation model for strategic forest management planning in the eastern Amazon. *Ecological Modelling* 173:335–354.

Pine, W.E. III, T.J. Kwak, and J.A. Rice. 2007. Modeling management scenarios and the effects of an introduced apex predator on a coastal riverine fish community. Transactions of the American Fisheries Society 136:105–120.

Rabinovich, J.E., M.J. Hernández, and J.L. Cajal. 1985. A simulation model for the management of vicuña populations. *Ecological Modelling* 30:275–295.

Sutton, T.M., K.A. Rose, and J.J. Ney. 2000. A model analysis of strategies for enhancing stocking success of landlocked striped bass populations. North American Journal of Fisheries Management 20:841–859.

Weclaw, P. and R.J. Hudson. 2004. Simulation of conservation and management of woodland caribou. *Ecological Modelling* 177:75–94.

Wiegand, T., J. Naves, T. Stephan, and A. Fernandez. 1998. Assessing the risk of extinction for the brown bear (*Ursus arctos*) in the Cordillera Cantabrica, Spain. *Ecological Monographs* 68:539–570.

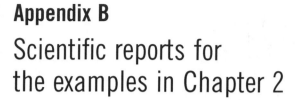

Appendix B
Scientific reports for the examples in Chapter 2

In this appendix, we communicate the results of our examples in a format suitable for a technical journal. The typical "methods" and "results" sections commonly found in scientific journals are awkward, since a simulation model is both a method and a result. We replace these sections with *overview of the model, model description, model evaluation*, and *simulation of* . . . We feel these sections better capture the process of model development and application. Other than changing the variable names, the models presented in these reports are identical to the final models from Chapter 2.

B.1 Effect of deforestation on rate of food harvest

Introduction

Recent deforestation has reduced the abundance of food items in the forest, perhaps by as much as 20%. This reduction in food availability may have a serious impact on a group of hunter-gatherers that live in the forest during winter. In this paper, we describe a model that simulates food harvest and the accumulation of food items by the hunter-gatherers over a 30-day period, assuming a 20% reduction in food items in the forest.

Background information

The group of hunter-gatherers usually hunts in the savanna until the end of October and then moves to the forest to spend the winter. Severe winter weather, which prevents hunting and gathering activities, typically arrives by the end of November. Thus, the group has only about a month to harvest the food items they need to survive the winter. There usually are about 100 food items in the forest at the beginning of November. For all practical purposes, food items are a nonrenewable resource during November since plants no longer produce fruits and both birth and natural death rates of animal populations are essentially zero. Historically, the group has had no trouble harvesting the 75 food items they need to survive the winter, although the group has told us the rate at which they can harvest food items decreases noticeably as abundance of food items decreases. The group estimates they can harvest about 10% of available food items per day. As a result of the recent deforestation, we estimate only about 80 food items will remain in the forest when the group arrives this fall. Thus, although there still appears to be enough food available, the group may not have enough time to harvest all they need before the severe weather arrives.

Overview of the model Our model represents the number of food items in the forest (*FI*), the harvest of food items (*H*) at a rate that depends on *FI*, and the total number of food items harvested by the hunter-gatherers (*TH*) (Fig. B.1).

Model description We have represented the model as a compartment model based on difference equations ($\Delta t = 1$ day). The number of food items in the forest decreases as they are harvested at a rate determined by the number of food items remaining in the forest. All harvested food items accumulate as they are stored for winter.

$$FI_{t+1} = FI_t - (H_t) \star \Delta t \qquad \text{(eq. B.1)}$$

$$TH_{t+1} = TH_t + (H_t) \star \Delta t \qquad \text{(eq. B.2)}$$

$$H_t = 0.1 \star FI_t \qquad \text{(eq. B.3)}$$

where FI_t and TH_t represent the number of food items in the forest and in the total harvest, respectively, at time t, and H_t represents the number of food items harvested during time t (number of items/day) (Fig. B1).

Model evaluation We evaluated the model by comparing simulated and observed decreases in number of food items in the forest over a 30-day period, assuming the initial number of food items was equal to that normally observed prior to deforestation (100). We also noted the number of days required for the simulated harvest to reach the 75 food items required by the group of hunter-gatherers for winter. Simulated and observed decreases in number

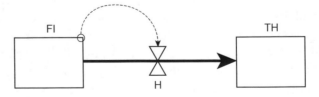

Fig. B.1 Conceptual model representing the food harvest from the forest by hunter-gatherers.

of food items in the forest were identical; in both cases, the number decreased exponentially from 100 to approximately 4 over the 30-day period (Fig. 2.3a). The number of days required for the simulated harvest to reach the 75 food items required by the group of hunter-gatherers was 15, which seems reasonable since historically the group has had no trouble harvesting the 75 items in 30 days.

We simulated the effect of deforestation on the rate of food gathering by initializing the number of food items in the forest at 80, rather than 100, and noting the number of days required to harvest a total of 75 food items. Simulation results indicate that the group could harvest 75 food items in 27 days (Fig. 2.3b). *Simulation of effect of deforestation on rate of food gathering*

Results of our simulations suggest that the group of hunter-gatherers should have enough time to gather the food they require for winter. However, the group may have only a few (we estimated 3) "extra" days before the expected arrival of severe weather. *Discussion*

We are concerned about the future of an animal population on a small, isolated island that is subject to hurricanes, particularly in view of a projected increase in hurricane frequency over the next few decades; historically, hurricanes have occurred about once every 10 years. In this paper, we describe a model that simulates probabilities of population extinction over the next 50 years assuming annual hurricane probabilities ranging from 10% to 50%.

B.2 Effect of hurricane frequency on probability of population extinction

Introduction

Available data indicate population size has fluctuated between 70 and 400 individuals over the past several decades. Annual per capita birth rate decreases linearly from a maximum of 0.7 when population size is 50 or lower to a minimum of 0.5 when population size reaches 400 (Fig. 2.6). Annual death rate usually is 50%, but increases to 99% during hurricane years. Currently, as a result of a hurricane last year, only 100 individuals per unit area remain on the island, the majority of whom survived the hurricane as eggs. *Background information*

Our model represents changes in population size (P) resulting from births (B) and deaths (D). The per capita birth rate (BR) depends on population size and per capita death rate (DR) depends on hurricanes (H) (Fig. B.2). *Overview of the model*

We have represented the model as a compartment model based on difference equations ($\Delta t = 1$ year). Population size changes as the net difference between births and deaths. *Model description*

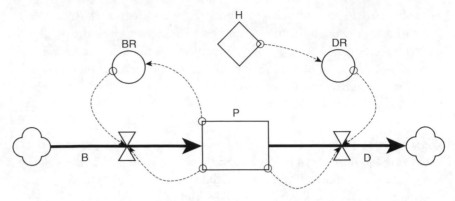

Fig. B.2 Conceptual model representing the dynamics of a hypothetical animal population whose mortality is affected by hurricanes.

$$P_{t+1} = P_t + (B_t - D_t) \star \Delta t \qquad \text{(eq. B.4)}$$

$$B_t = BR_t \star P_t \qquad \text{(eq. B.5)}$$

$$BR_t = 0.729 - 0.0005714 \star P_t \qquad \text{if } 50 \leq P_t \leq 400 \qquad \text{(eq. B.6)}$$

$$BR_t = 0.7 \qquad \text{if } P_t \leq 50 \qquad \text{(eq. B 7)}$$

$$BR_t = 0.5 \qquad \text{if } P_t > 400 \qquad \text{(eq. B.8)}$$

$$D_t = DR_t \star P_t \qquad \text{(eq. B.9)}$$

$$DR_t = 0.5 \qquad \text{if } H_t = 0 \qquad \text{(eq. B.10)}$$

$$DR_t = 0.99 \qquad \text{if } H_t = 1 \qquad \text{(eq. B.11)}$$

where P_t represents population size at time t (number of individuals), B_t and D_t represent the number of births and deaths, respectively, during time t (number of individuals/year), BR_t and DR_t represent per capita birth and death rates, respectively, during time t (number of individuals/individual – year), and H_t is a unit-less index representing the occurrence of a hurricane during year t (0=no hurricane, 1=hurricane). The value (0 or 1) of H_t is determined randomly, with the probability of a value of 1 being equal to the hurricane frequency that is being simulated.

Model evaluation   We evaluated the model by simulating population growth for 50 years without hurricanes. The simulated population grew from 100 to 400 individuals in approximately 25 years and stabilized at that level (Fig. 2.5b); this seems reasonable in view of what we know about historic population levels.

Simulation of hurricane effects on population extinction   We simulated population dynamics over the next 50 years assuming annual hurricane probabilities of 10, 20, 30, 40, and 50%. For each hurricane probability, we ran 100 replicate stochastic (Monte Carlo) simulations, each with the population initialized at 100 individuals. Simulation results suggest probabilities of population extinction during the next 50 years are less than 1% with annual hurricane probabilities of 10, 20, and 30%, but increase to 50% with annual hurricane probabilities of 40, and 50%, respectively (Tables 2.1B and C).

Results of our simulations suggest that annual hurricane probabilities as high as 30% pose no threat to the population, but as annual hurricane probabilities approach 40%, the probability of population extinction rises to levels that warrant close monitoring.

Discussion

B.3 Effect of stocking rate on forage and animal production

Introduction

Our neighbor plans to begin increasing the number of animals he puts on a common pasture we share; he plans to increase the size of his herd by 1 animal per year for the foreseeable future. In the past, we both have put 1 animal on the Commons at the beginning of January and have removed it at the end of December. We would like to add animals to our "herd" so we take advantage of our fair share of the available forage, but we are worried that both forage biomass and final weights, at harvest, of our animals will decrease over time. In this paper, we describe a model that simulates forage biomass and animal weights over the next 10 years assuming we (1) increase the number of animals we put on the Commons at the same rate our neighbor does, and (2) continue to put just 1 animal on the Commons each year.

Background information

Forage on the Commons fluctuates seasonally, but usually stays close to 1000 biomass units. Rate of forage growth depends on current forage biomass; growth rate decreases linearly from a maximum of 1 unit of biomass produced per unit of forage biomass per month to 0 as forage biomass increases from 200 to 1200 units. Forage is lost due to grazing by the animals; each animal requires 1 unit of forage biomass per unit of animal biomass per month. There also is a nongrazing forage loss due to trampling, which is negligible with only 2 animals on the Commons, but this loss may become important as the number of animals increases; each animal may trample about 2% of the forage each month. Animals are 2 months old and weigh about 10 biomass units when they are put on the Commons, and weigh about 90 biomass units when they are removed at 14 months of age. The maximum rate at which animals gain weight, assuming there is enough forage to meet all of their forage requirements, depends on their current weight; monthly weight gain decreases linearly from a maximum of 1.25 times current body weight to 0 as weight increases from 5 (the lower lethal body weight) to 100 units. Actual weight gain depends on the relative availability of forage; as the ratio of forage biomass available to forage biomass required decreases from 5 to 1, the proportion of maximum possible weight gain realized decreases from 1 to 0. Grazing animals normally metabolize 10% of their current biomass each month due to energy expenditures to meet maintenance requirements. Data from a grazing experiment conducted on three pastures similar to the Commons indicate the effect of moderate grazing pressure on forage growth (Fig. 2.17).

Overview of the model Our model consists of two submodels representing dynamics of (1) forage biomass (*F*) and (2) individual animal weight (*W*) (Fig. B.3). Forage biomass changes as the net difference between growth (*FG*) and grazing losses (*GL*) and nongrazing (trampling) losses (*NGL*). Forage growth rate (*FGR*) decreases as forage accumulates, nongrazing losses depend on number of animals on the Commons (*A*) and per capita nongrazing loss rate (*NGLR*), grazing losses depend on number of animals on the Commons and per capita forage requirements (*FR*), which, in turn, depend on individual animal weight. Animal weight gain (*WG*) depends on maximum rate of weight gain (*WGMR*), which decreases as animal weight increases, and a weight gain index (*WGI*), which depends on relative forage availability (*RFA*) (per capita forage availability). Weight loss depends on metabolic requirements (*MR*).

Model description We have represented the model as a compartment model based on difference equations ($\Delta t = 1$ month). Forage dynamics are calculated as:

$$F_{t+1} = F_t + (FG_t - GL_t - NGL_t) \star \Delta t \qquad \text{(eq. B.12)}$$

$$FG_t = FGR_t \star F_t \qquad \text{(eq. B.13)}$$

$$FGR_t = 1.2 - 0.001 \star F_t \quad \text{if } 200 < F_t \leq 1200 \qquad \text{(eq. B.14)}$$

$$FGR_t = 1.0 \qquad \text{if } F_t \leq 200 \qquad \text{(eq. B.15)}$$

$$FGR_t = 0 \qquad \text{if } F_t > 1200 \qquad \text{(eq. B.16)}$$

$$NGL_t = NGLR \star A_t \star F_t \qquad \text{(eq. B.17)}$$

$$GL_t = FR_t \star A_t \qquad \text{if } FR_t \star A_t \leq F_t \qquad \text{(eq. B.18)}$$

$$GL_t = F_t \qquad \text{if } FR_t \star A_t > F_t \qquad \text{(eq. B.19)}$$

where F_t represents number of units of forage on the Commons at time t (units of biomass), FG_t, GL_t, and NGL_t represent growth, grazing loss, and nongrazing loss, respectively, of forage on the Commons during time t (units of biomass/month). FGR_t represents forage growth rate (units of biomass produced/unit of forage biomass – month), $NGLR$ represents proportion of forage biomass lost due to trampling (0.02 of forage biomass lost/per animal – month), A_t represents total number of animals on the

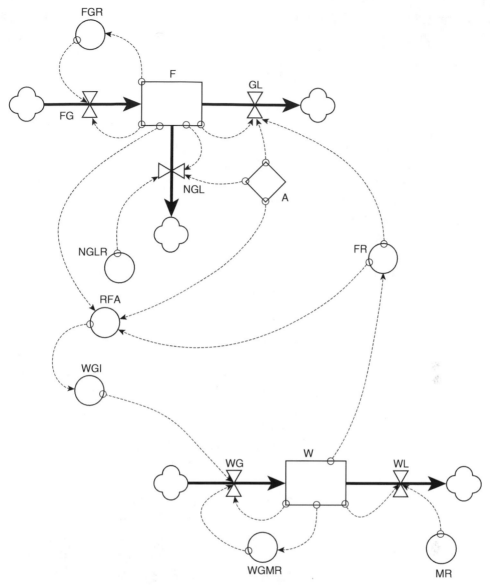

Fig. B.3 Conceptual model representing the forage and animal weight dynamics on the Commons.

Commons at time t, FR_t represents forage requirement per animal (1 unit of forage biomass/unit of animal weight – month).

Animal weight dynamics are calculated as:

$$W_{t+1} = W_t + (WG_t - WL_t) \star \Delta t \qquad \text{(eq. B.20)}$$

$$WG_t = WGMR_t \star W_t \star WGI_t \qquad \text{(eq. B.21)}$$

$$WGMR_t = 1.32 - 0.132 \star W_t \qquad \text{if } 5.0 < W_t \leq 100 \qquad \text{(eq. B.22)}$$

$$WGMR_t = 0 \qquad \text{if } W_t \leq 5 \qquad \text{(eq. B.23)}$$

$$WGMR_t = 0 \qquad \text{if } W_t > 100 \qquad \text{(eq. B.24)}$$

$$WGI_t = -0.25 + 0.25 \star RFA_t \qquad \text{if } 1.0 < RFA_t \le 5.0 \qquad \text{(eq. B.25)}$$

$$WGI_t = 0 \qquad \text{if } RFA_t \le 1 \qquad \text{(eq. B.26)}$$

$$WGI_t = 10 \qquad \text{if } RFA_t \ge 5.0 \qquad \text{(eq. B.27)}$$

$$RFA_t = F_t / (FR_t \star A_t) \qquad \text{(eq. B.28)}$$

$$WL_t = MR \star W_t \qquad \text{(eq. B.29)}$$

where W_t represents weight of the average animal on the Commons at time t (units of biomass), WG_t and WL_t represent weight gain and loss, respectively, of the average animal on the Commons during time t (units of biomass/month), $WGMR_t$ represents maximum rate of weight gain of an animal as a function of its current body weight (proportion of current body mass/month), RFA_t represents relative forage availability (a ratio of total forage to forage required by the total number of animals), WGI_t represents the proportional reduction in animal weight gain due to a shortage of available forage, and MR represents the constant proportion of body weight lost per month (0.1).

Model evaluation — We first evaluated the model by simulating a 12-month grazing experiment that had been conducted on three pastures similar to the Commons. Initial forage biomass in each pasture was 100 biomass units, and the pastures contained 0, 1, and 2 grazing animals, respectively, each initially weighing 10 biomass units. Simulated and observed forage dynamics were essentially identical (Figs 2.13, 2.19).

We also evaluated the model by comparing long-term (10-year) forage and animal weight dynamics with our expectations based on informal observations of the Commons. We initialized forage biomass at 1000 biomass units, placed 2 animals, each weighing 10 biomass units, on the Commons each January, and harvested them each December. Simulated and expected forage dynamics were essentially identical (Fig. 2.20a), and final animal weights also looked reasonable, although simulated animals were slightly heavier at harvest and actually attained their peak weight several months before harvest (Fig. 2.20b).

Finally, to determine the effect uncertainty in our estimate of trampling rate (nongrazing loss; proportion of forage biomass lost per animal per month) would have on our projections of forage biomass and animal weight dynamics, we conducted a sensitivity analysis in which we first reduced our estimate from 2% (baseline value) to zero, and then increased it to 5%; based on what we do know about the animals and the forage, it seems quite unlikely that trampling rate could be more than 5%. Results of simulations in which we maintained 2 animals on the Commons indicated forage dynamics were relatively unaffected by these changes (Fig. 2.21a), and animal weight dynamics were not affected at all (Fig. 2.21b). However, when we simulated 10 animals on the Commons each year, results indicated both forage biomass and animal weight dynamics were greatly affected (Fig. 2.22).

Based on results of our sensitivity analysis, it seems clear that (1) we should give high priority to gathering information that would allow us to estimate trampling rate with more confidence, and (2) in the meantime, it would be wise to make alternative projections of each of our scenarios using a range of trampling rates that appropriately reflects the degree of our uncertainty in its true value.

Simulated effect of stocking rate on forage and animal production

We simulated the effects of different stocking rates on forage biomass and animal weights over the next 10 years assuming we (1) increase the number of animals we put on the Commons at the same rate our neighbor does (add 1 additional animal each year), and (2) continue to put just 1 animal on the Commons each year (while our neighbor adds 1 additional animal each year). We simulated each of these two scenarios with each of three versions of the model, assuming nongrazing (trampling) losses of 2%, 0, and 5% per animal per month. For each simulation, we initialized forage biomass at 1000 units and animal weight at 10 units. Simulation results suggest that both forage biomass and individual animal weight at harvest decrease over time under each scenario, even assuming no trampling loss (Figs. 2.23, 2.24), although total animal biomass at harvest continues to increase for 2 or 3 years (Fig. 2.25).

An interesting trend in these initial results, which we first noticed during simulation of the grazing experiment, is that animals reach their maximum weight before harvest and then actually lose weight. The weight loss is not very noticeable with only 2 animals on the Commons (Figs 2.20b, 2.21b), but becomes increasingly noticeable as the number of animals increases (Figs 2.22b, 2.23b, 2.24b). This suggests that if we both harvested our animals earlier in the year, say, in June, we could harvest virtually the same animal biomass, and give the forage 6 months to recover. Our simulations of a June (rather than a December) harvest indicate that forage biomass can recover completely each year with the 6-month no grazing period, except at the highest trampling rate (Fig. 2.26a), and that total animal biomass harvested can be sustained over the 10-year period (Fig. 2.27). However, individual animal weights at harvest still decrease continually (Fig. 2.26b), and we notice that harvest weight in year 10 of the calculations assuming 5% trampling loss is 5.9 biomass units, which is becoming dangerously close to the lower lethal body weight (5 biomass units).

Discussion

Overgrazing appears inevitable if our neighbor adds an additional animal to his herd each year, even if we do not. Since total animal biomass at harvest seems likely to increase for the first 2 or 3 years under either scenario, this may erroneously lead our neighbor into thinking his plan is working quite well. If we are unable to convince our neighbor to forgo his plan to continue adding more animals each year, a promising alternative might be to advocate an early (June) harvest, which would have the dual benefit of harvesting animals closer to their peak weight and allowing the forage 6 months to recuperate from the increased grazing pressure.

Index

Note: page numbers in *italics* refer to figures; those in **bold** to tables.